**COMPUTER VISION
FOR
ROBOTIC SYSTEMS
An introduction**

COMPUTER VISION
FOR
ROBOTIC SYSTEMS
An introduction

MICHAEL C. FAIRHURST

Electronic Engineering Laboratories
University of Kent, Canterbury

PRENTICE HALL
New York * London * Toronto * Sydney * Tokyo

First published 1988 by
Prentice Hall International (UK) Ltd
66 Wood Lane End, Hemel Hempstead,
Hertfordshire, HP2 4RG
A division of
Simon & Schuster International Group

Printed and bound in Great Britain at
The University Press, Cambridge

Library of Congress Cataloging-in-Publication Data

Fairhurst, Michael, 1948–
    Computer vision for robotic systems.

    Bibliography: p.
    Includes index.
    1. Robot vision.   2. Computer vision.   I. Title.
TJ211.3.F35   1988   .   629.8′92   87-29246
ISBN 0-13-166927-3

*British Library Cataloguing in Publication Data
is available*

1   2   3   4   5   92   91   90   89   88

ISBN 0-13-166927-3
ISBN 0-13-166919-2 PBK

This life's five windows of the soul
Distorts the Heavens from pole to pole
And leads you to believe a lie
When you see with, not thro', the eye

WILLIAM BLAKE

Contents

Preface

Computer vision, by its very nature, is a subject that draws on a wide variety of traditional disciplines for its underlying formative principles, and in developing a distinctive methodology of its own it must first synthesize and subsequently build upon this elemental material. A book that aims to provide an introduction to this very important and potentially very productive field must therefore be rather carefully selective in the material it offers.

The literature is relatively well served with comprehensive and advanced texts on traditional image processing – the manipulation and transformation of pictorial data – and, perhaps to a somewhat lesser extent, on pattern recognition theory – the classification and categorization of object/entity descriptions. However, these disciplines are too often treated as essentially disjoint, whereas it is suggested that a more integrated view of these traditional areas might ultimately be to the greater benefit of both. Moreover, and most importantly, it is precisely these two areas that at a fundamental level may be seen as forming the foundations upon which the more complex image interpretation methodologies of computer vision 'practitioners' are constructed.

It is from these observations that the basic motivation for this book has sprung, as its principal aim is to introduce and integrate the topics and techniques that give the study of computer vision its particular character. Additionally, in order to give a significant, practical, realistic and topical context to the discussion, the principles to be dealt with are introduced within a framework that explores vision systems in industrial applications, specifically applications relating to robotics and automation, where the provision of visual input is desirable and potentially highly effective. The definition of what constitutes a robot system is here interpreted fairly broadly so as not to be unduly restrictive.

Specifically, then, the text aims to demonstrate that computer vision draws on a broad range of expertise, integrating this in a way which gives the subject its own identity. In particular it attempts to view computer vision in its broadest terms, identifying and introducing its basic principles, both to provide a self-contained discussion of the important topics and to offer a link to more advanced and specialized texts dealing in greater detail with particular areas. Where possible, techniques are illustrated with small-scale examples through which the reader can ensure that concepts are understood. Further examples for self-study will allow the reader to pursue and gather together the major ideas of each chapter, and will provide a course tutor with some useful material for student assessment.

A reasonably comprehensive set of references will point the reader to a range of further sources of information, including general textbooks at various levels, books and articles which extend particular topics in greater depth, and a variety of research papers which illuminate much more comprehensively some of the more important specialized topics. Hence, taken as a whole, the book should provide a sound introduction to this fascinating and important subject area, serving both as a first approach to its fundamental ideas and as a bridge between basic concepts and more advanced study. In this respect it is envisaged that this book could be used as an appropriate text for an introductory course in image processing, pattern analysis, computer vision or sensor-based robotic systems at degree, diploma or postgraduate level; as a supplementary text on a specialist course at an appropriate level in a relevant topic area; or as an introduction to image/pattern analysis for anyone new to the field as a student, practicing engineer, researcher or merely casual enquirer.

The mathematical content of the book has deliberately been kept to a minimum, to free the text from too many prerequisites. While no real programming knowledge is required in order to grasp the essential ideas introduced, it is envisaged that many students will want to implement for themselves some of the algorithms presented in order to explore them further and, indeed, such a course of action is strongly recommended.

The book begins with a chapter which both sets the overall scene and specifies the context within which the technical material is to be presented. In the following three chapters the emphasis is on the representation, transformation and analysis of images, moving from a basic infrastructure for image processing through the specification of some techniques for image enhancement to a discussion of processing operations which seek to extract structural information about the content of images. After a brief chapter which deals with some of the practicalities of computational architectures for image analysis tasks, the discussion moves more towards the classification/interpretation of images, beginning with an introduction to the basic ideas involved. This is followed by chapters which introduce the decision-theoretic techniques which underlie many interpretative procedures, a consideration of alternative strategies for image identification, and a detailed examination of one particular technique as a case study which, in adopting a specifically engineering approach to visual pattern recognition, reinforces the link to earlier material such as that which dealt with computational architectures. The reader is then led naturally to a brief concluding chapter which returns the discussion to the specific context of robotic systems with which the book began.

Consequently, ideas frequently recur, and cross-referencing should help the reader to see where topics easily and naturally interrelate. Similarly, in order to maintain a clear view of practicalities, Chapter 4 culminates with a discussion of how to make practical quantitative measurements of image characteristics and Chapter 9 with an example of a conceptually straight-forward but easily implemented system for pattern recognition. Thus the overall structure of the book reflects its fundamental intentions in a fairly obvious way.

In common with all those who have ever attempted to commit thoughts and ideas to print I am immeasurably indebted to more people than I can mention explicitly. However, it is appropriate that I express my thanks particularly to Glen Murray and, latterly, Andrew Binnie of Prentice Hall for their guidance and encouragement and, not surprisingly, to my wife Hilary for her support and tolerance of my very unpredictable working hours.

M.C.F.

1 ★ Robotics, computer vision and industrial processes

1.1 INTRODUCTION

This is a book primarily concerned with robotic systems. As will quickly become clear, however, in the present day such a statement does not necessarily convey a great deal about the content of the following chapters. This is because as the study of 'robots' has shifted from speculations of science fiction writers to the laboratories of researchers and, indeed, to the factory and other areas of practical realization, it has become increasingly apparent that, although the study of robotics is considered by many to be a separate academic discipline, it is more broadly seen as a synthesis of other established and emerging disciplines. Just as robotics represents a synthesis of ideas, techniques and principles, so also it is becoming more and more apparent that its accumulating body of knowledge, in common with all respectable scientific disciplines, is finding application in a variety of areas with differing requirements and differing emphases.

The fact that some topics of principal interest in this book are simultaneously claimed as the property of other disciplines points to one of the real challenges of the subject – that its sources are universally acknowledged as of importance and that its undisputed potential is yet to be fully realized.

This first chapter will deal with some of these background issues and identify the particular directions to be followed in subsequent discussions. It will also provide the framework within which the specific topics to be introduced later make sense as a whole. Since the book claims to be dealing initially with the subject of robotics in the broadest sense, we will be well advised to begin by defining some basic terms.

1.2 ROBOTS AND ROBOTIC SYSTEMS

Leaving aside the more lurid imaginings of popular literature, even within the community of scientists and technologists there is scope for divergence about precisely what constitutes a robot. For example, the Robot Institute of America was reported in 1983 as defining a robot in the following terms:

1

> A robot is a reprogrammable multifunction manipulator designed to move materials, parts, tools or specialized devices through variable programmed motions for the performance of a variety of tasks.
>
> (*Reported in:* Proc. IEEE, *January 1983*)

The problem with this definition is that it appears to hint at something rather too restrictive to encompass all current areas of activity, suggesting as it does the underlying notion of known, well defined and pre-specified operating characteristics. While many current industrial robot systems are concerned with precisely this sort of activity, much current research and development is summarized more readily in an alternative definition of the field attributed to Allen Newell:

> Robotics is the science of welding intelligence to energy – i.e. intelligent control of perceptually coordinated motion.
>
> (*Reported in:* Proc. IEEE, *January 1983*)

Although this definition itself has its deficiencies and, in particular, it might be argued that this view dismisses many of the currently installed and working systems – specifically those widely used to execute a pre-defined sequence of actions as a sort of programmable factory manipulator – it does seem to reflect the great breadth of the field rather better than the original definition.

This brief discussion demonstrates just why in a book of this sort it is necessary to identify fairly clearly the viewpoint from which the subject is approached. For our purposes, in dealing with the area of robotics, we shall tend more to the second definition than the first, though we shall restrict our study principally to one perceptual mode rather than the several which might be considered.

The second point to be made at this early stage is that we shall have in mind throughout the book that one of the principal application areas for systems which incorporate the techniques subsequently described is in the industrial field, and the examples chosen will largely reflect this.

In broad terms we may find it helpful to consider a robotic system as three interlinked components, as illustrated in Figure 1.1. The *processing* component is the heart of the system, controlling precisely what actions the system is to execute. In the simplest robotic system this means straight-forwardly translating a sequence of precisely specified instructions to an appropriate set of signals to initiate a number of (mechanically realized) actions on the external environment. In a well defined factory assembly process, for example, the robot might be instructed to pick up a component from a (specified) position on a transporter and attach it at a specified orientation to a part-assembled structure held in a fixed frame. In the simple system envisaged here these movements are entirely predetermined and the system is successful precisely because the operating environment

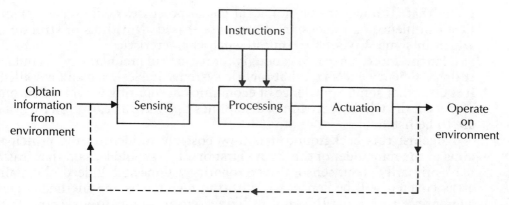

***Figure 1.1** An overview of some major components of sensor-based robotic systems.*

is constrained so that all the components are specifically located initially and all spatial relationships are likewise completely determined. The sequence of instructions, although initiated and interpreted by the processing unit would require considerable complexity in the *actuators* (a mechanical hand, for example, with several degrees of freedom) in order actually to execute these instructions.

Although in this particular example relatively little would be required from the *sensing* unit, since the process is organized so that the working environment is largely defined, in a more sophisticated system the sensing unit could be the key to the entire operation. For example, in some envisaged quality testing system, the sensing unit might be a TV camera positioned above a conveyor belt. The task of the system might be to perform, under the control of the processing unit, a visual inspection of manufactured components, allowing an acceptable product to pass down the line while discarding samples that in visual terms appear unsatisfactory in some way. In this case the sensing/processing units are of paramount importance, while the actuation unit could be relatively simple, perhaps only routing the moving samples in one of two possible directions after the visual inspection.

These two examples immediately point to the very broad range of the study of robotics; in addition to the obvious need to draw on the methods and principles of computer science/engineering (and Artificial Intelligence if this is regarded as a separate discipline) and electronic engineering, it is readily apparent that mechanical engineering, control engineering and so on are of considerable importance. Likewise, if a robotic system is to be sensor-driven—for example, using visual sensing as in the illustrative case described above—then it might be seen as advantageous to know a certain amount about some aspects of biological science, particularly if we acknowledge that most sensory processes, and vision most of all, are most efficiently and effectively carried out by biological organisms. It is perhaps likely that a knowledge of biological visual processes could have a significant influence on the design of computer-based algorithms or structures which in some way seek to imitate such characteristics.

Furthermore, since it is generally for reasons of productivity, manufacturing efficiency and so on that robotic systems are designed and installed, it is clear that some knowledge of economics/accounting would not be out of place if the whole spectrum of disciplines that have a bearing on robotics are to be listed.

Against this background it is now possible to identify the principal aims of the remainder of this book. First of all, it should be said that it is a book primarily about sensor-driven robotic systems and, indeed, the main areas covered will be concerned with the techniques and methodologies appropriate to interpreting and utilizing sensory input to a system. Furthermore, as the title of the book implies, we shall focus on the provision of visual input and investigate the sort of processing required if such information is to be productively utilized by a computer-based system. This will mean that relatively little will be said in any detail about the design of specific actuator systems, though many textbooks are available which cover this ground well.

Two further points need to be made. The first is that the book identifies two sub-disciplines which are often treated as separate areas for study and attempts to assign equal importance to each within an integrated framework. These two areas are respectively image processing and pattern recognition, and we shall treat these as two views of an operational domain with a common focus of attention. The second point to be made is that it is intended that the book should provide a broad-based introduction to the main issues within the general area of computer vision for robotic systems. This means that, while aiming to provide a thorough working introduction to the basic issues of significance, the book also seeks to bridge the gap between traditionally taught topics of computing/electronics and the abundance of texts that deal with these issues at a very specialized level.

We are now in a position to tackle some of the fundamentals. The remainder of this chapter sets the scene and provides a particular context – that of industrial automation engineering – for our study of topics in computer vision for robotic systems.

1.3   COMPUTER VISION AND ROBOTICS

In the context of our study, where we shall be concerned mainly with industrial/automation processes, it is possible to categorize some relevant areas of computer vision into two (clearly related) sub-areas. Their relation to computer vision and the broader context are illustrated in Figure 1.2. These two areas are:

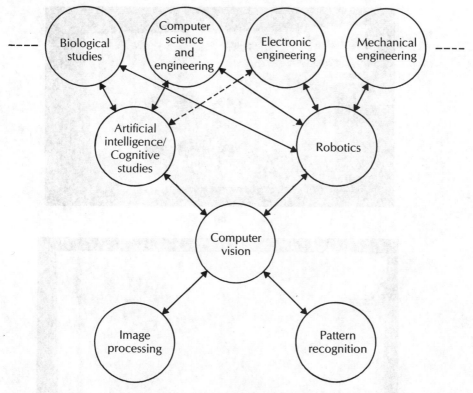

Figure 1.2 Interaction of various fields of study defining interests in computer vision.

1. *Image processing* This area deals principally with operations on images which aim to improve their 'quality' in some way, or to emphasize features of particular importance or relevance.
2. *Pattern recognition* This area is concerned with the identification or interpretation of the images. It aims to extract information (at a high level) about what the image is intended to convey.

   For example, Figure 1.3 shows an 'original' image together with the result of applying some processing (to be discussed in detail later) to improve its visual 'quality' with respect to criteria which could vary from case to case. The processed image offers the possibility of much easier extraction of information about its contents when viewed by a human observer, but also it is much more likely that an algorithm to automate the process of identification of the object of interest in the image will be effective when applied to the processed image rather than working on the original version. Of course, automatic identification requires further processing but some steps toward this are apparent in Figure 1.4, where some initial processing has extracted information about the general shape

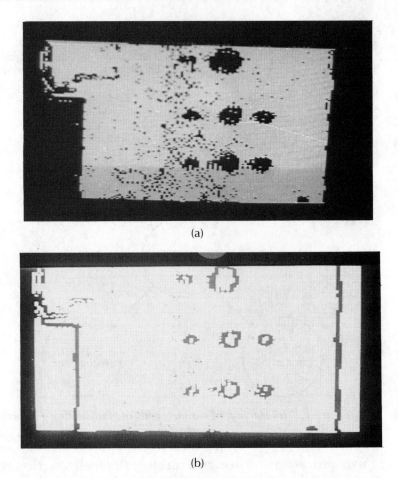

(a)

(b)

Figure 1.3 Operating on an image to improve its visual appearance.
(a) Image before processing. (b) Image after application of appropriate
processing.

characteristics of an imaged object. This might form the basis of some (initially perhaps rather crude) process where known objects are described in terms of characteristics that can subsequently be used in trying to categorize objects of unknown identity.

It is immediately clear that there are very close links between the two broad areas of image processing and pattern recognition. Nevertheless, although it is advantageous to keep in mind these links, it is true that each area generates its own methodologies and techniques, and these will be explored in some detail in later chapters.

Of course, computer vision is a large and expanding subject area, and it is useful to point out that pattern recognition in the present context will be taken to mean the identification/interpretation of images that generally

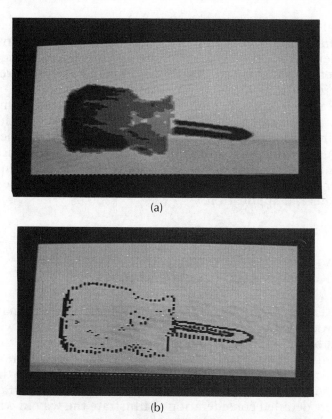

(a)

(b)

Figure 1.4 Operating on an image to extract information about object outline and basic shape. (a) Image before processing. (b) Image after appropriate processing.

contain a single object or entity or a particular structure. Consequently we shall not be concerned necessarily with highly complex visual scenes (which might be involved in more general-purpose vision-based robot systems) except to the extent that some initial processing can often localize the area of an image to be examined, and so make some formal pattern recognition techniques more widely applicable.

These ideas will develop and the implied constraints become more apparent as we proceed. For the present, however, we need note only that the techniques to be introduced form the underlying theory of all automatic visual processing systems, and that a good grasp of the principles introduced here will support a study of any computer vision system.

One final point can be made. Since we are working toward the *automation* of vision processing, some thought needs to be given to the actual implementation – realization in physical form – of techniques and algorithms which can be described in theory. Although it is natural that we should immediately look to the conventional digital computer as the ideal general-purpose tool providing the means for implementation, we shall examine this question in more detail later: one of the interesting features to emerge from the study of problems in computer vision is the way in which the special nature of the task domain has stimulated a consideration of appropriate supporting computational structures.

Finally in this chapter we shall address the fundamental issues that define the major application area of interest. In doing this, not only shall we gain an insight into the underlying motivation for such studies, but we shall also be able to identify the specific problems that dictate the direction our subsequent investigation needs to follow. Consequently, the next section will look briefly at the general area of vision-based robotic systems for industrial application.

1.4   AN OVERVIEW OF INDUSTRIAL APPLICATIONS FOR VISION-CONTROLLED ROBOTIC SYSTEMS

Potential application areas for vision-driven automated systems are numerous. Each brings its own particular problems which must be resolved by a system designer if successful operation is to be achieved but, generally speaking, applications can be categorized broadly according to the processing requirements they impose. To illustrate this we will briefly describe a number of such areas of application and then take one case study for more detailed consideration to illustrate the way in which the nature of the task domain influences the design requirements of a robotic system.

Example 1: Object identification

A vision system may be required to identify objects or items. For example, objects – or perhaps more likely, identification markings – may be imaged and recognized as part of a sorting process in, say, an automated ware-housing environment.

Example 2: Picking/placing tasks

Vision control may be used to guide the actions of a mechanical manipulator to effect physical movement of items. The precise situation can vary considerably. For example, objects may be picked from a moving conveyor belt or, as is also often the case in practice, lifted from storage bins which provide a common and efficient means of factory storage and transport.

An example might be the transfer of samples from the output of one sub-process in a manufacturing chain to the start of the next sub-process, or the acquisition of some finished item for packing. It is clear that bin-picking tasks are generally inherently more difficult to automate than tasks represented by Example 1 because of the basic lack of organization (and the consequent lack of reliable constraint information) of the objects concerned.

(Note also that many practical tasks could involve a combination of the above two task categories, where identification might influence the subsequent acquisition procedure.)

Example 3: Visual guidance

A visual capability might be used to guide the movement of some manipulator in the execution of its task. This is particularly helpful where a relatively passive process, involving the accurate specification of relative positions and movements, would be unlikely to succeed.

An example might occur in some automated assembly process where, say, two components are fitted together by matching at an appropriate position and orientation. For example, it is easy to envisage how one component might be held fixed while the other is maneuvered by a mechanical hand, the movements of which are directed by visual feedback from a visual sensing system.

Example 4: Trajectory control

This is really a type of operation in the category described by Example 3 above, though it is sufficiently typical of a whole range of industrial processes to merit its own category.

In this type of situation, visual control may be used to guide the movement of some mechanical manipulator along some trajectory which cannot be specified in absolute terms, but only in relation to a set of criteria which are most easily determined by on-line visual sensing.

An example in this category might be the guidance of a cutting tool which is to describe an operational trajectory across a metal surface defined in relation to some previously assembled, surface-mounted structure.

Example 5: Visual inspection

Inspection by visual means is a very obvious and potentially highly beneficial application area, which can be used as a powerful tool in automating quality control procedures and in obtaining specific quantitative measurements of important parameters in a manufacturing process.

Inspection criteria can range from a simple pass/fail decision about the general appearance of an object or item to a diagnostic procedure which may be used to recover or control errors, and can be implemented with varying degrees of required accuracy.

An example might be a process that inspects a package to ensure that the requisite number of items of a given range of shapes and sizes is present. As an example of a situation where higher accuracy is of primary importance, automatic visual inspection might be used to derive quantitative measurements ensuring that some set of components are manufactured to within certain allowed tolerance levels.

It should be apparent even from these introductory examples that the requirements of a computer vision system could vary markedly both across a particular defined task category and even more so between different problem categories. To consider this point in rather more detail we will now examine one particular case study in somewhat greater depth.

1.5   A VISION SYSTEM FOR AUTOMATIC CODE READING – A CASE STUDY

Let us consider a hypothetical problem area in which a system is to be designed to inspect labels on cartons visually and read them automatically. Labels are to be typewritten (but may be from differing sources) and contain a six-character alphanumeric code, and to be attached to the carton at a fixed predetermined position. It is assumed that an imaging device consisting of an appropriate camera and controller is available.

Although this situation, which might be a part of the development of, say, an automatic warehousing system, is not specified in exhaustive detail here it nevertheless provides a concrete example from which to extract some points of importance which illustrate the general nature of the problems with which the designer of a vision-driven system is typically faced.

There are, first of all, some essentially high-level issues to be considered:

1. *Locate object of interest*   The constraint identified in the problem specification that indicates that labels are attached to cartons at a known position is extremely helpful, as it means that minimal effort is required to locate the area of specific interest to the vision system in relation to the rest of the visual environment.

   Nothing is said here about the physical movement of the cartons, and we shall assume that some suitable arrangements are made to position the carton appropriately relative to the imaging sensor.

   In relation to the area of interest – the label in this case – it is still necessary to deal with some questions at a relatively high level; for example the following.
2. *Lighting conditions*   An appropriate level and distribution of illumination will be necessary in order to obtain an image of acceptable quality for the extraction of information of relevance for the identification task.

Inadequate lighting will, at best, degrade image quality and impose greater demands on the processing module and, at worst, might make character identification impossible or unacceptably unreliable.

3. *Segmentation* If the alphanumeric code were to be imaged and treated as a single entity, i.e. as a single individual 'object', the problem of identification would be potentially very difficult, since each code example would be associated directly with its own specific identity category. Assuming no constraints on character grouping, there are up to 36^6 (2 176 782 336) different codes and the processing system would therefore need to choose among this huge number of categories in reading the code.

   If, on the other hand, we precede the code identification process by 'segmenting' the acquired image or, in other words, breaking it up into its constituent components – in this case into six separate characters – the processing will require six steps, but each step will involve a choice between only 36 different categories.

4. *Image quality* The image obtained may be of inherently poor quality. For example, perhaps because of the nature of the operating environment, the labels become dirty or smudged, or the alignment of a label may not be particularly good, causing skewing of the characters. Likewise, poor image quality may occur because of poor imaging parameters at the sensing stage, or could be related to problems of lighting and so on.

   Whatever the cause, it may be necessary to operate on the 'raw' data in order to improve the chances that subsequent processing will result in the correct reading of the data.

5. *Processing speed* The physical environmental in which the system operates will inevitably impose some constraints on the processor. In particular, it will be necessary for the system to carry out the identification of the label code at a sufficiently high speed to keep pace with the arrival of subsequently presented cartons and, indeed, within the bounds of the time available to activate any further routing or transporting mechanisms.

   This will obviously be linked with questions relating to the actual computation involved in the code identification process, but is clearly likely to be of some importance in a real-world task domain such as that outlined in the current example.

In addition to the considerations already introduced, there are a number of other factors, often concerning rather lower level detail, which have implications for the system designer. Among these are the following:

1. *Feature extraction* It is necessary to decide upon the inherent characteristics of different object classes (i.e. alphanumeric characters in this case) on which an algorithm for code identification is to be based. Not

only that, but decisions will have to be made about how such features are to be computed from the raw image data.

For example, most common imaging devices will automatically generate, at the raw data level, a set of discrete measurements of, say, image intensity which could be used directly as the basis for identification. On the other hand, a typical television camera would generate a very large number of such picture points and it may be better to try to obtain information about, perhaps, some global or geometric attributes of the characters involved such as their overall shape, component line segments, and so on.

2. *Identification* The specification of the actual algorithm used for the character identification is generally nontrivial. For one thing, it is necessary to decide how to combine the information available about the chosen features of interest in order to compute the identity of a particular character.

It is clear that, since data is generated from an unspecified number of different sources, this is not likely to be an easy task and, indeed, it is unlikely that a completely foolproof or error-free system can be designed in practice. Note also that, with a code that is segmented as already described, individual character-level error rates will be cumulative with respect to the performance of the system on the whole code.

3. *Utilization of context* It may be helpful to consider whether there is any information inherent in the structure of the label code that can either resolve specific ambiguities in identification or generally reduce the number of errors that might occur.

For example, if it were known that the code is always to begin with a letter, the identification problem at the first code position would reduce in complexity, and we would not need to worry about the possibility of confusion between, for instance, a B and an 8, which otherwise might prove rather troublesome.

It is clear that in addressing these points there is much interaction between the areas involved, and decisions in one area will necessarily have a bearing on decisions in another. For example, a decision to adopt a particular approach to character identification will have a direct influence on the sort of operations carried out on the raw data. If taking an approach that tries to match intensity values at relative spatial locations between an imaged character and prestored samples of all possible character types, it would be necessary to try to normalize the data—for example, in terms of character size or position within the limits of the available viewing window—to have any reasonable chance of success.

The example chosen illustrates how a particular situation can give rise both to questions of a general value and those of a highly specific and task-oriented nature. This particular case study is also quite a useful one to consider in some detail not only because it provides a good illustration of the nature of a typical task for vision-based systems, but also because the issues raised will all be covered in the remainder of this book. By the end of Chapter 9, therefore, it should be possible for the reader to return to the current example with a reasonable idea of how to specify a system that broadly meets its requirements.

1.6 EXAMPLES FOR SELF-ASSESSMENT

1.1 Describe, giving specific examples where appropriate, the range of typical industrial applications for robotic systems and suggest how such systems might usefully be categorized.

Discuss in some detail the requirements for an automated system that uses visual information to locate and identify workpieces on a conveyor belt. Outline the additional considerations that might be necessary if the components were to be picked from storage bins rather than located on a belt.

1.2 Carry out a preliminary analysis, along the lines of the case study given in the text, for the following situation. A system is to be designed to carry out a visual inspection of samples of motor car brake shoes in a manufacturing process. The samples appear one at a time at some specified inspection point on a conveyor belt and the result of the inspection process is the generation of a signal to activate a mechanical sorting mechanism to direct the sample to one of two different destinations.

1.3 Compile a list of sensory mechanisms that might be used to obtain controlling information for a robotic system. Identify the main advantages and disadvantages of each and give an example in each case of an application where a particular sensory mode might be most effective.

2 ★ An operational infrastructure for image processing

2.1 INTRODUCTION

Whether we are primarily interested in performing operations on images that make them more easily interpreted by a human observer, or whether our main concern is to automate (or, more specifically in the present context, 'computerize') the manipulation or interpretation of images, a fundamental requirement is to generate an appropriate method of representing image data. The method chosen should serve two specific purposes. It should encapsulate information which defines the important characteristics of the image and it should provide a basis for convenient and efficient processing by computer.

In this chapter we will consider the basic problem of image representation and the implications of that representation in relating an 'object' to its 'image', and then examine some practical questions that arise in dealing with image data. Later chapters will explore image processing operations in more detail as we move toward questions of practical system implementation and the interpretation of images.

2.2 IMAGE REPRESENTATION

If we consider the idea of an image from the simplest point of view we could regard an image as a two-dimensional function, where the value of the function $f(x,y)$ at spatial coordinates (x,y) in the $x-y$ plane defines a measure of light intensity or brightness at that point.

For our purposes, where ease of computer-based processing is a primary requirement, we shall be concerned with *digital* images, where we accept a suitable approximation of the function $f(x,y)$ in return for the convenience in representation and subsequent processing. A digital image is an image which has been approximated in two ways, corresponding to *spatial digitization* and *amplitude digitization*, as follows.

Spatial digitization Sometimes referred to as image sampling, this involves representing the notional original continuous image function as an array of specific samples at discrete points within the two-

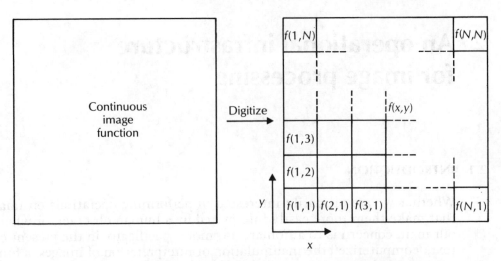

Figure 2.1 Definition of a pixel reference grid in image digitization.

dimensional frame of reference, as illustrated in Figure 2.1. In general, the spatially digitized image consists of $N \times N$ equally distributed samples, where each discrete point on the array is identified as a picture element or *pixel*. Of course, other array dimensions offering different perspectives may be appropriate in particular cases, but we shall assume a square array unless otherwise indicated.

Amplitude digitization   Each pixel in the array has to encode the local image intensity and this, too, is approximated by assigning to it a value corresponding to one of a fixed number of quantized levels. An image intensity level is designated a *gray level* and the full range of intensity levels available in a particular image is referred to as a *gray scale*. The number of gray levels within the gray scale is often chosen (for computational convenience) to be a power of two, and hence any pixel in the $N \times N$ array is found to have an intensity level, *g*, where

$$0 \leqslant g \leqslant 2^{l} - 1$$

and *l* is an integer.

2.3 BINARIZED IMAGES

A special case of amplitude digitization occurs when the gray scale consists of just two possible levels (i.e. $l = 1$), as the resulting image consists of an array of points each of which is either completely black or completely white. Such an image, comprising an array of two-valued pixels, is known as a *binarized* image.

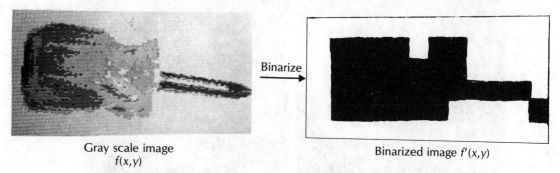

Gray scale image
$f(x,y)$

Binarize

Binarized image $f'(x,y)$

Figure 2.2 Approximation of a gray scale image through binarization. This also illustrates the effect of low-resolution spatial quantization.

The generation of a binarized image from a more general gray-scale image is exceedingly simple in practice, and requires only the introduction of an appropriate thresholding level to partition the image into pixels with one of just two values.

Hence, if

$f(x,y)$ = intensity value at coordinates (x,y) in the image

and

T = thresholding value

then, as illustrated in Figure 2.2, the image represented by $f(x,y)$ is binarized by applying the rule:

If $f(x,y) \geq T$ then $f'(x,y) = 1$

If $f(x,y) < T$ then $f'(x,y) = 0$

where $f'(x,y)$ denotes the binarized version of $f(x,y)$.

In the example we have adopted the convention that black is identified with $f'(x,y) = 1$ and white with $f'(x,y) = 0$, though this association is arbitrary.

It is clear that binarized images are simple to generate, store and manipulate, since each pixel is associated with a single bit of information. As we shall discover in more detail in later chapters, provided that we retain enough information for subsequent processing requirements, there are many advantages in working with binarized images where possible in practical situations.

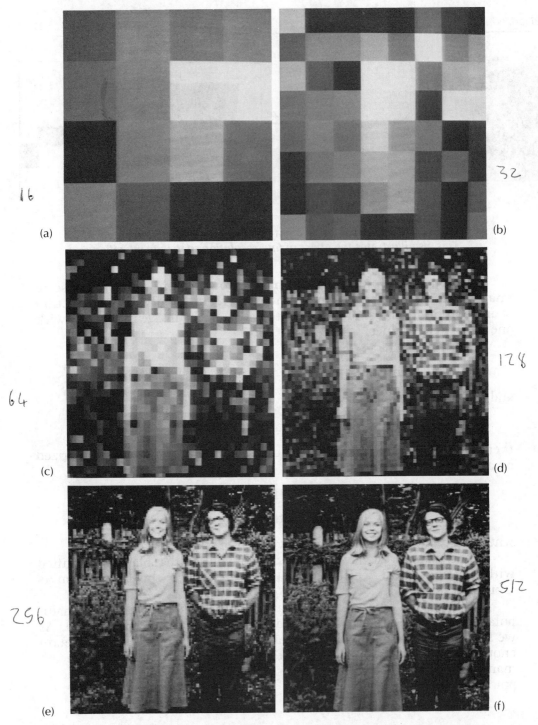

16

32

(a)

(b)

64

128

(c)

(d)

256

512

(e)

(f)

Figure 2.3 Illustration of the visual effect of varying spatial resolution. N increases from 16 to 512 between (a) and (f). (From Ballard, D.H. and Brown, C.M. Computer Vision, Prentice Hall, 1982.)

2.4 CHOICE OF DIGITIZATION PARAMETERS

Even at this early stage of discussion it is possible to identify two very important practical points. The first is that the two digitization parameters described above – the number of pixels and the number of gray scale levels – will determine the degree of *resolution* of an image. To put it another way, the relationship between these two parameters will, in a very practical sense, determine the extent of the low-level detail the image can convey. Figures 2.3 and 2.4 show some examples of varying spatial and amplitude digitization and the effects produced, and it is clear that a suitable choice must be made in practice in relation to both the sort of information that must be preserved in the image and the use for which the image is intended. This point was implicitly made in the binarization example referred to in Figure 2.2, where it is useful to note the visual effect of digitizing an image with a relatively low-resolution spatial quantization when the precise shape of the object of interest is not very well preserved.

The second important point is that the choice of spatial and amplitude digitization parameters will determine the amount of storage required to hold and process an image. In practice this can be an important consideration, and there is clearly a trade-off between increasing resolution to preserve detail and increasingly demanding storage requirements. The balance between the two competing effects can be determined only in relation to a specific situation.

In general, then, the number of bits required for storage of a digital image is simply derived as

$$\text{Number of bits} = P \times l$$

where P = number of pixels in digitized array (N^2 using the regular square array described above), and 2^l = number of gray scale levels available.

Again, in a practical situation, this may represent a lower bound on the storage requirements, since it may be wise further to trade storage for a computationally convenient organization of pixel data to avoid crossing memory–word boundaries, and consequently the value determined above is only an approximate guide to hardware requirements.

2.5 ARRAY TESSELATION

In digitized format, images are represented as an array of discrete points – the pixels – where intensity values fall within the gray scale chosen for the image. It is almost intuitive to assume that these pixels will be organized spatially to form a square ($N \times N$) array and, indeed, this is the most common arrangement. However, it is not the only possible configuration (or

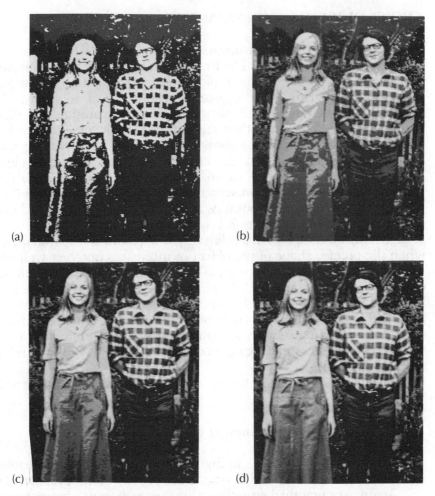

(a)   (b)

(c)   (d)

*Figure 2.4 Illustration of the visual effect of varying amplitude
resolution. l increases from 1 to 8 between (a) and (d). (From Ballard, D.H.
and Brown, C.M. Computer Vision, Prentice Hall, 1982.)*

tesselation) and the advantages and disadvantages of such a choice ought to
be considered.

For the purposes of discussion, let us consider a very simple, low-
resolution, binarized image (see Figure 2.5) of an 'object' interpreted as a
line drawing as shown, where the line is of single pixel thickness.

The first problem to consider is that of connectedness. In other words
it is necessary to define the conditions under which two pixels are connected. For example, it would be useful to know whether two black pixel
points in a binarized image are part of a continuous line or not. This may
seem to be a perfectly straightforward question to which the answer is self-
evident. However, two possibilities exist which we will examine in turn.

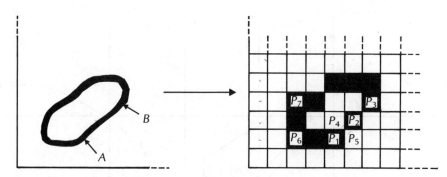

Figure 2.5

Use the 8-C rule According to the 8-C (8-connected) rule a pixel is 'connected' to any of its eight immediate neighbors, both those in the N, S, E and W directions, and also those that touch at diagonals (i.e. NE, SE, SW and NW). If we use this definition of connectivity, the pixels labeled P_1 and P_2 in the binarized image of Figure 2.5 are connected, as are pixels P_2 and P_3, the pixel group P_1, P_2, P_3 collectively representing the binarized version of the straight line connecting the points A and B in the original.

A problem now arises if we apply the same rule to pixels P_4 and P_5, since these, by virtue of the connection established between P_1 and P_2 should *not* be connected. There is consequently ambiguity in the definition of connectedness, where different rules apply to black pixels from those applying to white pixels.

Use the 4-C rule According to this rule, a pixel may only be considered to connect with its four immediate neighbours in the N, S, E and W directions – that is, where pixels meet at edges rather than diagonal points. It is quickly seen, however, that this only turns the problem round, for now P_1 and P_2 are, according to the definition, disconnected. Unfortunately, so are P_4 and P_5 if the same rule is applied and consequently, once again, it is necessary to invoke different definitions of connectivity for black and white points.

In practice, the difficulty of defining connectedness on a square pixel array does not cause a particularly severe problem, though it is important to be aware of its existence, so that an appropriate interpretation of boundaries and areas may be made.

2.5.1 Distance measurements

In extracting information from images it is often useful to be able to measure distances between various points in an image. If Euclidean distance is

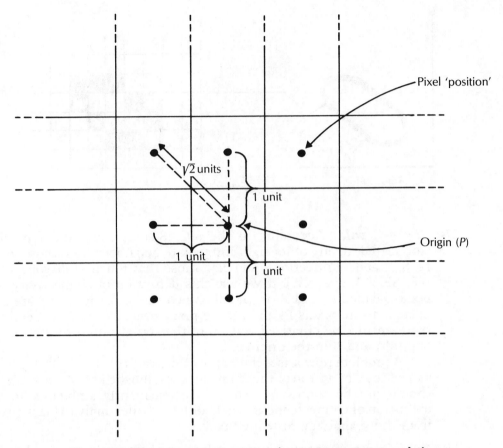

Figure 2.6 Distance measurements using a square array tesselation.

the metric adopted, as is often the case, it is again important to be aware of an inconsistency when using a square array tesselation.

This is illustrated in Figure 2.6, and we assume that the precise 'position' of a pixel is located by reference to the centre of the small area covered by the pixel. It is seen that the eight immediate neighbors of an arbitrary 'origin' pixel P are found not to be equidistant from this reference point. For example, in Figure 2.5 pixels P_6 and P_7 are equidistant from P_1 if a measurement is made on the basis of a simple pixel count (three pixel distance), but the Euclidean distances P_1 to P_7 and P_1 to P_6 are different. This can present a problem which might need attention in some practical applications.

All of this suggests that alternative array tesselations might have advantages in some respects. The most common alternative generally considered is a hexagonal configuration of pixels such as that shown in Figure 2.7. It will be apparent that in this format there is no ambiguity in relation to the question of pixel connectivity, as adjacency at any of the six sides

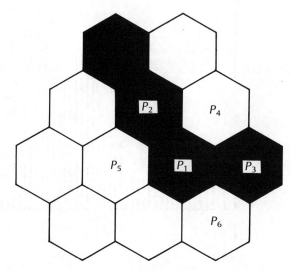

Figure 2.7

fulfills the requirement for two points to be connected, irrespective of whether we are looking at black or white pixels. For example, it is evident that pixels P_1 and P_2 are connected, leaving P_4 and P_5 disconnected. Likewise, P_1 and P_3 are connected, defining P_4 and P_6 to be disconnected in an entirely consistent way.

Similarly, the difficulties caused by differences in inter-pixel distances are not apparent, since examination of any arbitrary pixel point (P_1, for example) will reveal it to be equidistant from all of its six immediate neighbors.

Despite these advantages, however, the hexagonal array has a severe disadvantage in that its configuration destroys the natural orthogonal horizontal/vertical axes inherently fundamental to many images encountered in practical situations. The square array remains the most popular and widely used geometric configuration of pixels, and this will be adopted as the standard for the remainder of this text.

2.6 THE GRAY LEVEL HISTOGRAM

As indicated in Chapter 1, image processing is often concerned with improving the 'quality' of images in some respect – either to make them more easily interpretable by a human observer or to facilitate the extraction of important features by some automated/computerized system. Clearly, in order to do this it will often be necessary to identify groups of related pixels for specific transformation (for example, to identify an 'edge'

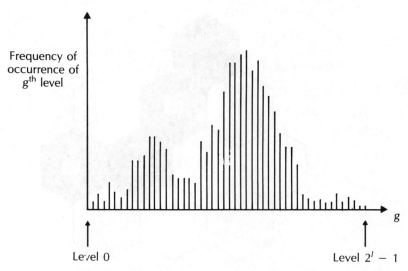

Frequency of
occurrence of
g^{th} level

Level 0   Level $2^l - 1$

Figure 2.8 Representation of image characteristics using the gray level
histogram.

between black/white areas), but a surprising amount of information about
the nature of an image may nevertheless be obtained from a much more
generalized view of its constituent pixel information. Hence, before considering more specifically how inherent image information might be
used to effect particular transformations it is useful to look at one of the
simplest and most general measures of the 'content' of an image – the gray
level histogram.

   If an image is scanned pixel by pixel across its entire area it is possible
to measure the frequency with which each of the quantized levels in the
available gray scale occurs. If this information is expressed in the form of a
histogram distribution, as illustrated in Figure 2.8, this is said to be the
gray level histogram of a particular image.

   Now although this histogram gives information about the overall distribution of gray levels that occur in the image, in representing that information in this way, all information about the precise spatial attributes and
inter-pixel relationships within the image has been destroyed. Furthermore, the relationship between a particular image and its gray level histogram is inherently nonreciprocal. That is to say, while a particular image
will generate a gray level histogram uniquely, it is not possible to work
backwards and uniquely derive a specific image from the information in a
gray level histogram. For this reason the gray level histogram does not
permit us to compute any properties specific (in a spatial or structural
sense) to the image from which it is derived. Such a representation is
nevertheless extremely valuable for two main reasons:

1. Despite its generality, the gray level histogram can give information

about the fundamental nature of the image and certain aspects of its quality.
2. It can provide a basis for defining some useful operations for image transformation.

The most useful information provided by the distribution depicted in the histogram relates to the degree of *contrast* within the image – the balance between the different possible levels. Reference to Figure 2.9 will illustrate some of the important characteristics displayed by the histogram. Figure 2.9(a) shows a histogram corresponding to an image within which the gray levels occurring are roughly evenly distributed across the available range of the gray scale used. Such an image makes good use of the gray scale provided, with the expectation that, within the limits of the spatial resolution adopted, detail will be discernible and maximal subtle gradations of shading observable within the constraints imposed by the gray scale chosen.

If we consider Figure 2.9(b) by comparison, it is evident that the corresponding image in this case would have considerably poorer contrast and consequent potential loss of detail, since the available range of gray scale values is comparatively poorly utilized. Retaining the same approximate shape while pushing the peaks towards the higher or lower intensity ends of the gray scale range would likewise correspond to poor contrast images at the predominantly dark or light end of the gray scale respectively.

A related problem occurs when, even with a generally reasonable distribution of gray scale values, an artificial peak occurs at either extreme of the gray scale range. (An example is shown in Figure 2.9(c).) In this case the form of the histogram suggests that the imaging device has artificially compressed the darker regions of the image into too small a number of gray scale levels, giving very poor contrast in the darker region with a consequent loss of detail which could be significant.

The gray level histogram can therefore, even at this stage, give some useful indications about the nature of the image from which it is derived. Identification of potential difficulties in this way offers the opportunity to adjust the operating parameters of the imaging system so as to re-digitize the image (if the observer has control over the imaging operation to this extent) or, as we shall see, to consider applying some appropriate transformation to the image that will modify its gray level histogram in a suitable way. The latter alternative will be considered further in Chapter 3 when some basic image processing operations are introduced.

Of course, we are not in every application concerned overmuch about fine detail within an image and, indeed, not every image necessarily possesses a great deal of fine detail. Nevertheless, even in such situations the gray level histogram can provide a useful source of information. Consider, by way of an example, a situation in which we are dealing with a highly constrained type of image which depicts an approximately uniform dark

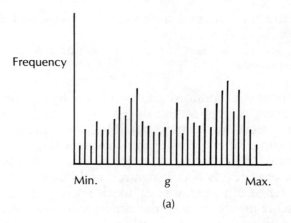

(a)

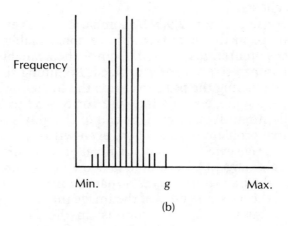

(b)

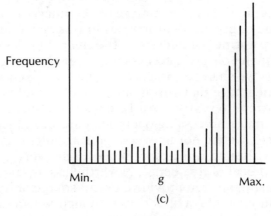

(c)

Figure 2.9

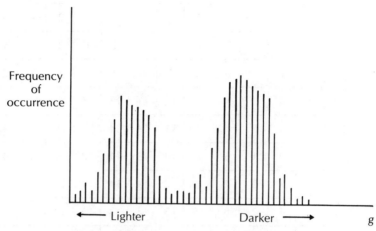

Figure 2.10 A bimodal gray level histogram.

object on an approximately uniform light background (images of simple components on a conveyor belt, where lighting has been carefully controlled, could fall into this category, for example).

In this type of environment it could be the case that we are aiming to work with binarized images for reasons of economy and processing speed. Typically the imaging device might produce as raw data a gray scale image with a distribution of quantized levels as represented in the sketch of Figure 2.10. In order to binarize this image it is necessary to choose a thresholding value as described earlier, and it is clear that the choice of this value will determine, to a large extent, the character of the binarized image. The gray level histogram can be very useful in making an appropriate choice. In the typical case illustrated, the peaks of this bimodal distribution show a group of pixels of approximately uniform intensity predominantly in the dark area (the object) and a group of pixels in the predominantly light area (the background). This distribution clearly suggests that a choice of thresholding level in the intensity region between the peaks ought to be optimal in producing a binarized representation of the original raw image. An inappropriate choice of thresholding value could, of course, have a significant effect on the shape and size of the object area in the binarized image, and it is worth considering a short example to illustrate this point.

Figure 2.11 represents a digitized image of 144 (12 × 12) pixels and 16 gray levels where, for convenience, the gray level at each pixel is numerically represented. We might easily imagine that this is an image of some object obtained in some unspecified manufacturing or inspection process, where the object shape and size is of greater importance than any subtle variation in surface appearance or internal detail. To binarize this image to emphasize just such characteristics of shape and size a binarization threshold, T, must be chosen.

1	3	5	2	1	0	0	4	1	2	1	0
2	1	3	2	2	1	3	2	1	2	5	3
1	1	2	3	2	1	0	2	3	1	3	2
2	3	2	4	9	10	11	11	13	14	11	5
1	3	2	4	5	11	14	14	13	15	9	4
1	0	1	5	7	14	13	11	12	11	10	4
1	1	0	7	8	12	14	14	11	12	4	1
3	4	2	8	15	15	13	13	13	12	13	9
0	1	3	8	14	15	15	13	11	8	3	3
1	1	2	9	13	12	15	14	11	12	4	2
1	3	10	13	14	15	14	15	14	12	10	7
1	1	2	5	6	12	10	5	6	2	1	2

Figure 2.11 Numerical specification of a gray scale image.

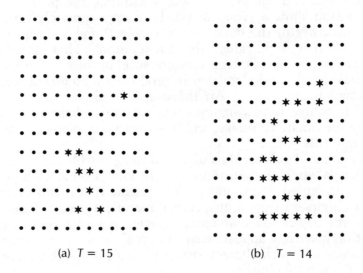

(a) *T* = 15   (b) *T* = 14

(c) *T* = 11

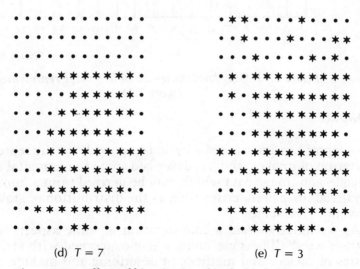

(d) *T* = 7   (e) *T* = 3

Figure 2.12 Effect of binarization threshold choice on image characteristics.

Figure 2.12 illustrates the effect on the resulting object representation of choosing different values for *T* in the binarization operation. It is clear from a study of this example that very different shape and size characteristics are generated by choosing different threshold values, and it is left to the reader to consider, perhaps by examining the gray level histogram for this image (shown in Figure 2.13), what the optimal choice might be in this case.

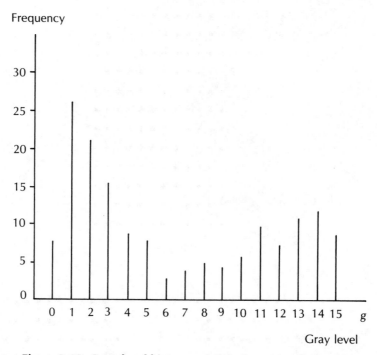

*Figure 2.13 Gray level histogram of the image represented in
Figure 2.11.*

2.7 SUMMARY

This chapter has introduced a basic framework for the representation and description of images, and has described some fundamental characteristics of images and image data which may be derived from a knowledge of only generalized image properties such as the distribution of gray levels within the overall gray scale.

As we move toward a consideration of some aspects of image interpretation we shall become increasingly concerned with structural characteristics of images and methods of describing and making measurements upon the objects represented. In the next chapter, however, we shall consider some particular image processing operations which are principally intended to improve the quality of images rather than to allow automatic extraction of information from them.

2.8 EXAMPLES FOR SELF-ASSESSMENT

2.1   Discuss how the requirements of a computer-based image processing system would depend on the resolution of the images to be processed. For example, compile a table to show how image storage requirements depend on spatial and amplitude digitization parameters.

2.2 Create some sample two-dimensional gray-scale images for experimental investigation (note that only relatively low resolution is required for illustration) and binarize these using different binarization thresholds.

Compare the gray-scale and binarized versions with respect to such properties as storage requirements, visual appearance and so on.

2.3 Discuss the principal practical difficulties which might be encountered in using an 8-bit processor to process data representing images on a 128×128 pixel matrix and a gray scale consisting of eight different levels.

3 ★ Basic image processing operations

3.1 INTRODUCTION

We have seen in Chapter 2 how images may be represented in a form suitable for computer-based manipulation. Furthermore, in the discussion of the gray level histogram we have seen how a conceptually straight-forward representation of even simple image characteristics can convey very useful information about the inherent nature and 'quality' of the image itself.

In many, and indeed most, tasks that involve image manipulation, the raw data obtained directly from the sensing/acquisition device will give a relatively poor-quality representation of the object or scene of interest. This problem, as we have seen, can arise because the image acquisition device is badly calibrated, but might also arise because of noise in the system, the fact that the image was obtained under poor lighting con-ditions, the existence of a less than ideal or unfriendly operating environ-ment, erroneous information introduced during data transmission, or for a variety of other reasons.

Such poor-quality data, however, may often be processed so as to improve its acceptability to an observer or to emphasize certain task-specific features of interest, or so that it is more suited to some automatic means of interpretation. In order to do this it is generally necessary to transform the raw image data by some appropriate means, when the trans-formed image may be expected to be of a higher quality in terms of the criteria of current importance. The derivation and refinement of appro-priate and generally applicable image processing algorithms has increas-ingly occupied researchers over the years, and this chapter will introduce some of the principles involved and describe some illustrative common processing procedures.

Two broad categories of operations may be identified. In the first category, a prime objective of the image transformation is to improve the visual quality of the raw image as it appears to a human observer (though, of course, such transformations are usually also valuable in applications involving automatic/machine interpretation). Such algorithms may be said to be concerned in a general sense with *image enhancement*. In the second broad group can be found image transformations concerned primarily with operations intended to modify data in a way that aids interpretation in some more specific way – perhaps, for example, by normalizing an object within an image according to some topological criteria. Although it will become clear that these two categories are by no means always distinct or mutually exclusive, it may be helpful initially to have them both in mind. The aims of image processing operations may be understood, then, as to effect a transformation which converts an image function $f(x,y)$ – usually the raw image data – to a modified function $f'(x,y)$, where $f'(x,y)$ has certain desirable properties that $f(x,y)$ did not have.

3.2   IMAGE ENHANCEMENT PROCESSING

The simplest way to transform the visual appearance of an image is to arrange that any given pixel value in the raw image $f(x,y)$, irrespective of the actual spatial location of the pixel in question, is mapped directly to a new value in the transformed version $f'(x,y)$. The transformation can then be carried out using an algorithm which is very easy to implement on a pixel-by-pixel basis. Since this type of technique is concerned solely with individual pixel values rather than any topological or spatial information, our only concern in these cases is with the type of information conveyed by the gray level histogram. For this reason such operations are often defined in terms of histogram effects and are referred to as *histogram modification techniques*.

We have already met one such operation; the binarization process introduced in Chapter 2 was a very straightforward pixel-based operation. We examined each pixel, compared its value against a threshold, and mapped its value to one of just two possible transformed values according to a very simple rule.

We may represent this type of operation more generally by reference to Figure 3.1. The function Φ, representing an image transformation, operates by mapping each gray level g in the image $f(x,y)$ – which has a gray level distribution given by $h(g)$ – to a corresponding value g' in the new image $f'(x,y)$, with a corresponding distribution of gray levels described by $h'(g')$.

In the case of the binarization operation; this function Φ, of course, would be just a step function, but other functions are clearly appropriate in other cases. We shall consider two further examples to illustrate this type of image enhancement operation.

3.2.1   Linear gray level shifting

Let us consider a hypothetical situation in which a raw image of an object produces the gray level histogram shown in Figure 3.2(a). Following our previous convention, this might typically represent a relatively light object against a fairly uniformly dark background. Now suppose that in this case the image is to be displayed on some arbitrary display device which suffers

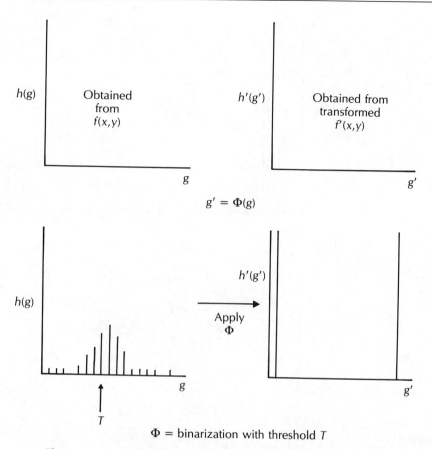

$$g' = \Phi(g)$$

Φ = binarization with threshold *T*

Figure 3.1 *Basic principle of image transformation, illustrated by reference to the image binarization case.*

from unwanted or unpredictable nonlinearities for gray scale levels in the area g_1 to g_2. If we were to utilize the raw data alone this would lead to distortions of the object of interest in the image displayed which might be unacceptable for some particular application. It is easy to see, however, that a histogram modification technique can be invoked which will cause a uniform shift in all gray scale levels, moving the peak of the gray level histogram distribution along the g-axis so as to allow display without distortion of its important features. The simple form of Φ shown in Figure 3.2(b) is all that is required, as this applies an equal linear shift (θ) to all gray level values within the image.

3.2.2 Enhancement through histogram flattening

The examples described above could only loosely be regarded as enhancing an image, though in the immediately preceding case, of course, the effect to

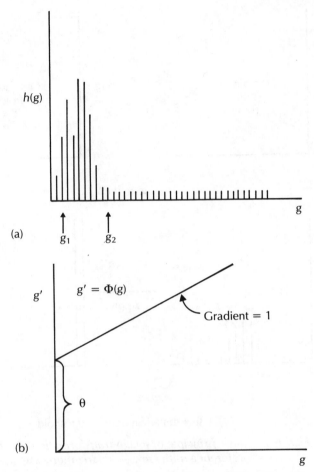

Figure 3.2 *Representation of a linear shifting operation.*

the observer of the entire system (including the particular malfunctioning display) would be to enhance his perception of the object of interest. More generally, however, enhancement would be expected to involve mappings of pixel values which are nonlinear in some way. This typically involves modifying the contrast within the image (the *relative* values of different gray levels) to change its appearance.

An example of a function that will impose certain changes in contrast is the transformation function Φ given by

$$g'_k = \frac{(2^l - 1)}{N^2} \sum_{i=0,k} h(g_i)$$

where g_n and g'_n denote the nth gray level for $0 \leqslant n \leqslant 2^l - 1$, and N^2 is the number of pixels in the image array, assuming a square tesselation.

If we consider by way of an example an image processing system working with images, each with 16 possible gray scale levels, on a 128×128
point square array, then in terms of the changes induced in the form of the
gray level histogram the effect of applying the transformation described
above can be illustrated by reference to Figure 3.3. Figure 3.3(a) shows an
arbitrary gray level histogram in which there is a bunching of gray level
values in the middle of the gray scale. The application of the histogram
modification function is to map pixel values in a way that produces a
transformed image with the gray level histogram shown in Figure 3.3(b).
This histogram is seen to be generally flatter in its appearance (perhaps
more easily seen if the approximate histogram envelope is visualized), with
the gray level values more evenly spread across the entire range of the gray
scale. The effect in terms of image characteristics of applying this function
Φ would be to 'stretch' or 'shrink' the contrast of the image in areas of high
and low pixel density respectively, with the result that detail may emerge
from an otherwise fairly indistinct image.

The function described is, in fact, recognizable as the so-called *cumulative distribution function*, describing the cumulative pixel population at
successive levels in the gray scale, and is a common histogram modification function. In principle, the application of this particular modification
function should transform the raw histogram to a form which has a perfectly uniform distribution of pixels across all possible gray levels (i.e. is
perfectly flat in appearance). The deviations from this ideal which occur in
practical situations such as that illustrated is caused by the fact that we are
dealing with discrete rather than continuous distribution functions, and
the consequent quantization approximations cause the result to be less
than perfect. Nevertheless the end-product of a histogram equalization
transformation can be quite impressive. Figure 3.4 shows an example of
this transformation operating on a real image to give an idea of its visual
effect.

To demonstrate the underlying theory of this approach it is more convenient to consider the gray level histogram as a continuous probability
density function (Figure 3.5(a)) which is to be transformed in such a way as
to generate the uniform gray level distribution shown in Figure 3.5(b), by
using the transform function Φ which relates level g in the original histogram to a corresponding level g' in the transformed version. Φ then has the
effect of mapping gray levels of the original image in the small interval dg
into a corresponding small interval dg' in the transformed version, where

$$h(g)\mathrm{d}g = h'(g')\mathrm{d}g'$$

From Figure 3.5(b) it is clear, however, that

$$h'(g') = \frac{N^2}{(2^I - 1)}$$

and hence

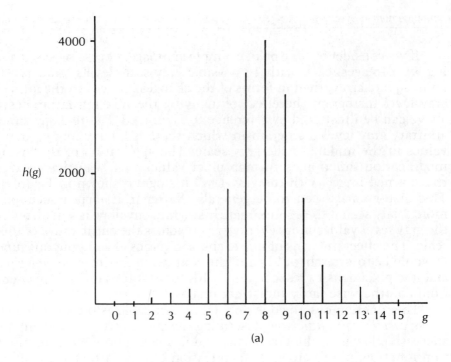

(a)

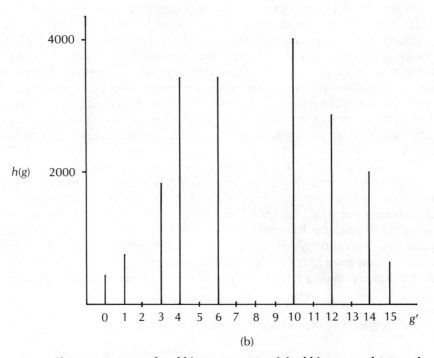

(b)

Figure 3.3 A gray level histogram: (a) original histogram, h(g), and (b) transformed histogram, h'(g'), after the application of a histogram flattening operation.

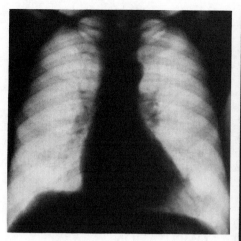 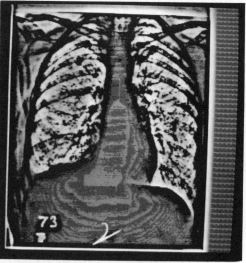

Figure 3.4 A practical example of contrast enhancement through histogram flattening: (a) original image, (b) transformed image. (From Ballard, D.H. and Brown, C.M. Computer Vision, Prentice-Hall, 1982.)

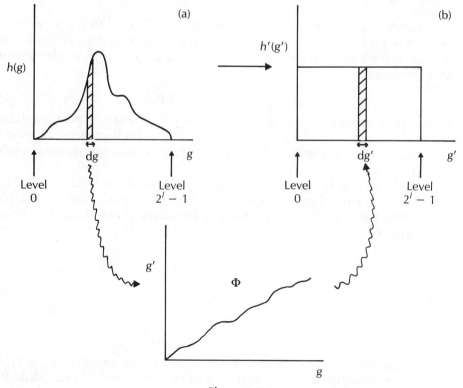

Figure 3.5

$$\frac{N^2}{(2^I - 1)} \, \mathrm{d}g' = h(g)\mathrm{d}g$$

Thus, we may write

$$g' = \frac{(2^I - 1)}{N^2} \int h(g)\mathrm{d}g$$

which is the continuous form of the discrete cumulative distribution function derived above.

3.3 PROCESSING BASED ON LOCAL AREA COMPUTATION

The processing operations described in the preceding sections rely on the assumption that each pixel value in the original image maps directly and uniquely to a new value in the transformed image. Another class of very useful processing algorithms, however, performs a rather more subtle computation, where a transformed pixel value is calculated as a function of a group of pixel values in some specified spatial location in the original image.

In other words, we apply a transform function, Φ, not to a generalized gray level g as such, but rather to a particular localized pixel such that, in transforming a raw image $f(x,y)$ to its processed form $f'(x,y)$, the operation Φ is given by

$$f'(x_i,y_j) = \Phi(\{P_{ij}\}) \qquad \text{for all } i,j$$

where $\{P_{ij}\}$ represents the gray level values of a set of pixels related to the (x_i,y_j)th pixel in the original image, and which in general would be expected to refer to pixels in the immediate localized spatial neighborhood of pixel (x_i,y_j).

In Figure 3.6, for example, we have identified a reference pixel (x_i,y_j) and denoted the pixel values of immediately surrounding neighbors, for convenience, as $P_0, P_1, \ldots P_7$. The function Φ could then be selected according to, say,

$$f'(x_i,y_j) = \frac{1}{4} \sum (P_1 + P_3 + P_5 + P_7)$$

or

$$f'(x_i,y_j) = \frac{1}{8} \sum (P_0 + P_1 + P_2 + P_3 + P_4 + P_5 + P_6 + P_7)$$

In the first case the transformed pixel would assume the gray level value corresponding to the average value of its 4-C neighbors, while in the

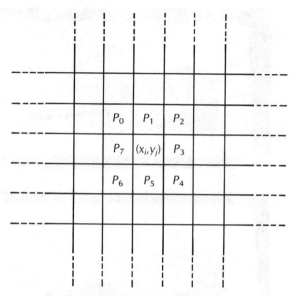

Figure 3.6

second all neighbors according to the 8-C rule would be used in determining the transformed pixel value.

If we now apply transformations such as these on a pixel-by-pixel basis across the entire image (making special arrangements, of course, for pixels at the extreme edges of the image for which there are fewer neighbors), then it is clear that we would generate a new image in which gray scale transitions have been smoothed out by the averaging process. *Smoothing algorithms* such as these are particularly useful in diminishing the effects of discontinuities and noise in an image. Also, of course, the algorithm can be extended and modified to accommodate different situations and requirements, perhaps by considering a larger number of neighbors in the set $\{P_{ij}\}$ – weighting the average over a larger radius from the pixel of interest – or by weighting the contribution to the summation of different neighbors in a different way. An example of the application of a smoothing operation is illustrated in the transformation shown in Figure 3.7.

It is convenient to consider this type of local area calculation at every pixel to achieve the transformation process in terms of a window, defining the neighborhood of interest, which moves across the image, centering on each pixel in turn, as illustrated in Figure 3.8 (with the 4-C neighborhood identified).

Used in conjunction with a thresholding operation, a smoothing function operating on binarized images can be quite effective in 'cleaning up' images and in removing spurious noise points, to produce images in which objects of interest are well defined in respect to their general structural,

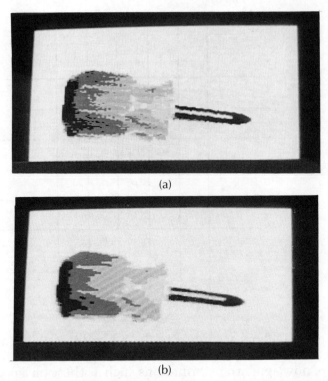

(a)

(b)

**Figure 3.7 Example of the application of a smoothing algorithm
(a) before, and (b) after smoothing.**

shape or geometrical features. In such operations Φ may be specified such
that:

$$
f(x_i, y_j) = \begin{cases} 1 & \text{if} \quad \sum_{\text{all elements}} P_{ij} \geq \theta \\[2em] 0 & \text{if} \quad \sum_{\text{all elements}} P_{ij} < \theta \end{cases}
$$

where θ $(0 \leq \theta <$ order of $\{P_{ij}\})$ is some chosen threshold.

Some examples of the application of this algorithm on images of binarized alphanumeric characters are shown in Figure 3.9, where an 8-C connection rule is assumed.

It should be noted that, as is the case with almost all image processing
algorithms, some care must be exercised in choosing the parameters of a
smoothing algorithm. Smoothing out discontinuities can most valuable,
but not if in the process features of situation-specific significance are
destroyed. For example, it is quite possible to devise an algorithm that
would fill in the spurious void in the left-hand limb of the A character in
Figure 3.10(a), but we must be aware that such a procedure might also,

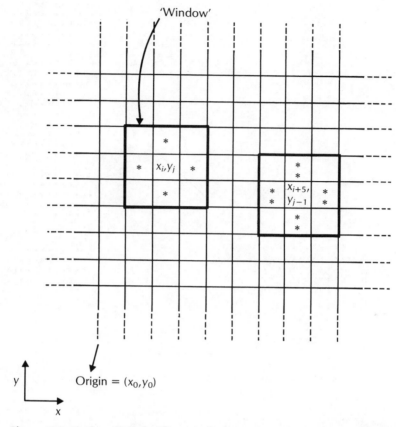

Figure 3.8 *Basic principle of a 'windowing' operation.* $*$ *= elements of* $\{P_{i,j}\}$, $**$ *= elements of* $\{P_{i+5,j-1}\}$.

if applied to the F character of Figure 3.10(b), cause a closing up of the correctly placed gap at its right-hand extremity, turning the character into a false P.

3.4 PREPROCESSING FOR NORMALIZATION

The algorithms described above all aim to improve the 'quality' – to use an ill-defined term – of an image in some way. It is frequently the case, however, that an automatic image interpretation or feature extraction procedure requires the implementation of a data transformation which operates, not so much to change the inherent image characteristics, but rather to normalize certain properties, perhaps for reference purposes or to allow some matching process to take place.

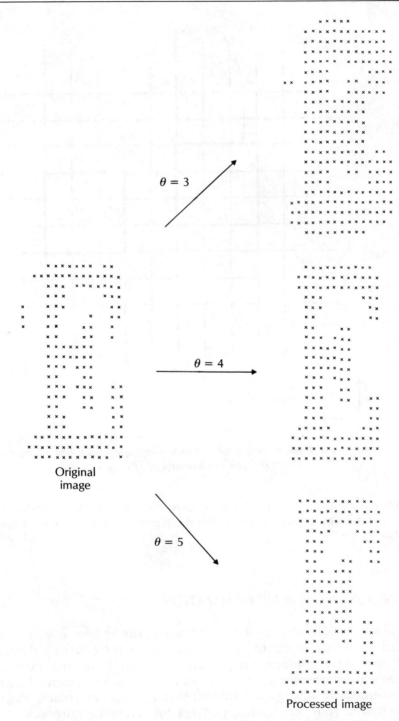

$\theta = 3$

$\theta = 4$

$\theta = 5$

Original
image

Processed image

Figure 3.9 Effect of smoothing/thresholding on a binarized image.

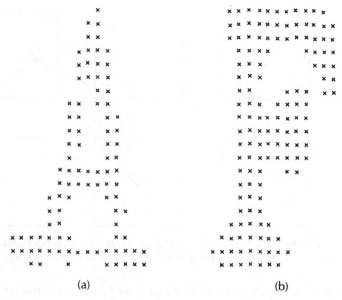

(a)   (b)

Figure 3.10

For example, Figure 3.11 shows some examples of an alphanumeric character (we shall assume binarized images for convenience), in this case the letter T, located within the boundaries of a reference grid defining the image matrix. If we are designing a system that will automatically read, for example, carton labels, it is important that all these sample characters are recognized as belonging to the same pattern class. If we work directly with the unprocessed image data on an $N \times N$ pixel array as in the example, then each of these images (seen in terms of an absolute spatial distribution of black/white points within the boundaries of the image matrix) will be vastly different, even though the 'object' of interest in each case has common properties which are significant in this particular application. Unless we adopt some interpretation algorithm that utilizes features that are independent of characteristics such as position, size or orientation, it will be necessary to introduce some data transformations to normalize the images with respect to these particular characteristics prior to the interpretation phase of processing. We shall consider two such normalization operations by way of example.

3.4.1 Position normalization

The simplest way to normalize an object in an image with respect to its position within the image boundaries is to implement a relabeling of axes, where coordinates are translated so as to align the extremities of the object

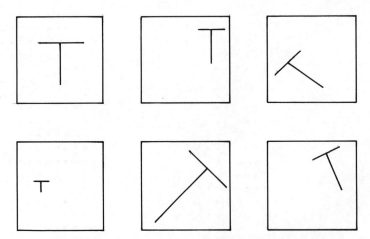

Figure 3.11 Illustration of typical variability in object appearance,
suggesting the need for normalization processing.

of interest at some point within the overall image matrix (see Figure 3.12 for example).

An alternative approach, which can take a better account of the geometric features of an object, is to 'center' the object of interest within the image matrix. This requires the determination of the horizontal and vertical axes of gravity, giving the coordinates of the center of gravity, $(x^\star, y^\star)$, of the object within the existing image matrix. These coordinates may be obtained from the calculation given by

$$x^\star = \frac{\displaystyle\sum_{\text{all } y} \sum_{\text{all } x} x \cdot f(x,y)}{\displaystyle\sum_{\text{all } y} \sum_{\text{all } x} f(x,y)}$$

$$y^\star = \frac{\displaystyle\sum_{\text{all } y} \sum_{\text{all } x} y \cdot f(x,y)}{\displaystyle\sum_{\text{all } y} \sum_{\text{all } x} f(x,y)}$$

The axes may then be relabeled to center the object at the point $(x^\star, y^\star)$, to produce the transformed (centered) image. An example of the application of this algorithm is shown in Figure 3.13.

3.4.2 Rotation

To normalize data for feature extraction purposes, or perhaps to obtain a 'best fit' in some pattern-matching operation, it may be desirable to rotate

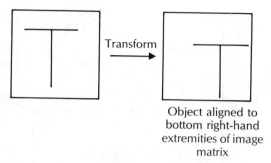

Figure 3.12 Simple axial alignment for position normalization.

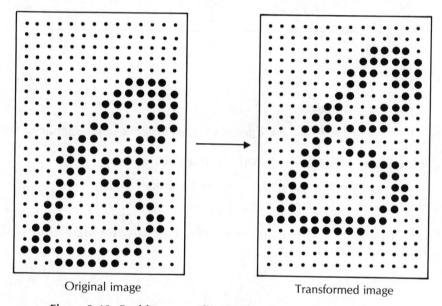

Original image   Transformed image

Figure 3.13 Position normalization based on 'centre of gravity' calculation.

an object in an image. Rotation algorithms generally allow the user to perform such a rotation through some arbitrarily specified angle about, say, the axes of the center of gravity of the object, obtained as derived above.

This transformation again requires only a recalculation of the co-ordinates of the pixels defining the object. For example, to rotate an object counterclockwise through an angle of $\theta°$ requires the general translation $(x,y) \rightarrow (x',y')$ shown in Figure 3.14. Simple geometry shows that we must effect a coordinate translation such that

$$x' = x\cos\theta - y\sin\theta$$

$$y' = x\sin\theta + y\cos\theta$$

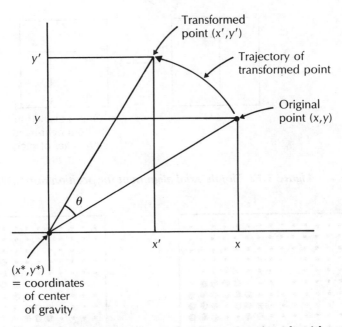

Figure 3.14 Basic geometry of an image rotation algorithm.

which may be expressed in the general form

$$\begin{bmatrix} x' \\ y' \end{bmatrix} = \begin{bmatrix} \cos\theta & -\sin\theta \\ \sin\theta & \cos\theta \end{bmatrix} \begin{bmatrix} x \\ y \end{bmatrix}$$

An example of the operation of this particular algorithm is shown in Figure 3.15.

3.5 PATTERN THINNING

For a variety of reasons depending on particular image acquisition processes and imaging resolution, ambient lighting conditions and variability, noise accumulation, and so on, a binarized image may often contain an object of interest whose inherent structure is to some degree obscured by the fact that lines/limbs are thicker than necessary (i.e. of multiple pixel width). An example is shown in Figure 3.16(a). In some applications it can be extremely useful to apply a transformation that reduces all lines or limbs to single pixel thickness while preserving the overall geometric properties of the object in question. The image processing literature has, over the years, accumulated a large number of such line-thinning algorithms, but many common approaches utilize a technique based on an algorithm developed by Sherman, which may be illustrated as follows.

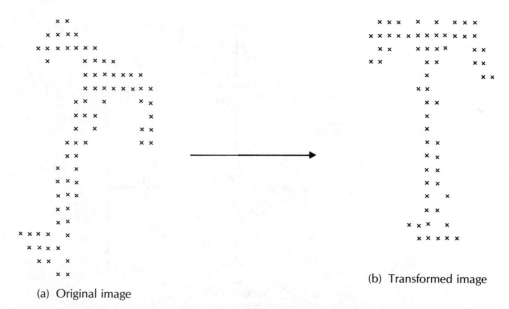

(a) Original image

(b) Transformed image

Figure 3.15 Example of the effect of a rotational transformation.

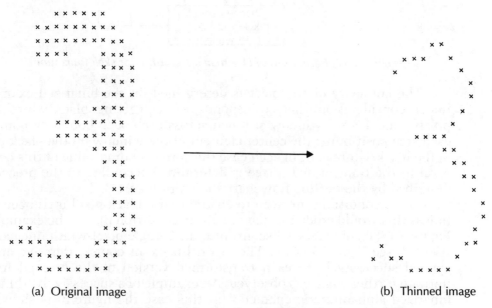

(a) Original image

(b) Thinned image

Figure 3.16 Practical example of a skeletonization procedure.

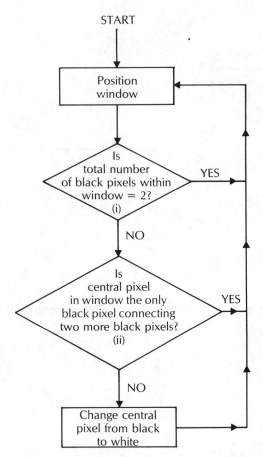

Figure 3.17 Basic form of Sherman's algorithm for skeletonization.

The boundary of an object is determined (in the binarized examples under consideration this corresponds to identifying black/white edge pixels), and a 3 × 3 window scanned across each such boundary point, at each step positioning the center element of the window on the black pixel of the black/white pair. In each case the corresponding value of this center pixel in the transformed image is determined according to the procedure described by the outline flowchart shown in Figure 3.17.

These conditions are seen to ensure that a black pixel is 'rubbed out' unless this would erode the end of a limb (condition (i), see the example of Figure 3.18(a)), or create a discontinuity in a single-pixel-width line (condition (ii), see Figure 3.18(b)). The algorithm is, of course, iterative and is applied successively to each transformed version of the original image until no further change is observed. An example of a successively 'thinned' binarized alphanumeric character (in this case the figure 3) is shown in Figure 3.16(b), in comparison with its original unthinned form.

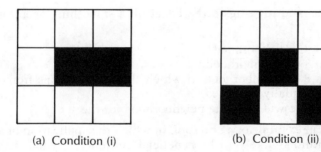

(a) Condition (i)   (b) Condition (ii)

Figure 3.18

3.6 SUMMARY

We have seen in this chapter how various transformations or preprocessing algorithms may be applied to directly acquired image data to change its appearance, emphasize certain characteristics, or modify the data in some other way that reduces subsequent dependency on features which are not relevant to some specified further processing. As we shall discover later (in the chapters dealing with the techniques available for pattern recognition), operations such as those described may not only be particularly influential in determining the ultimate performance of procedures and algorithms for automatic interpretation of pictures or images, but may also determine the acceptability and intelligibility to a human observer of picture data which has been acquired by a computer or other special-purpose hardware system.

3.7 EXAMPLES FOR SELF-ASSESSMENT

3.1 Investigate the effect of image subtraction – the pixel-by-pixel subtraction of gray scale values at corresponding points in two images. Describe a possible application area where this technique might be useful.

3.2 Discuss how the gray level histogram of an image may be used in the following cases:

(a) the calibration of an imaging device,
(b) the determination of a suitable thresholding level for image binarization,
(c) the derivation of a quantitative measure of the physical characteristics of an object within an image (state any assumptions made).

Illustrate your answer with an example in each case.

3.3 Determine the center of gravity of some sample patterns using the method described in the text.

Explain any drawbacks this method has and suggest and evaluate an alternative form of 'center of gravity' measure which might have advantages in some cases.

3.4   Define, and investigate the effect of, a smoothing operator which incorporates:

  (a) a 4-C neighborhood,
  (b) an 8-C neighborhood,
  (c) an 8-C neighborhood in which the contributions from the set $\{P_{ij}\}$ are not equally weighted,
  (d) a more wide-ranging neighborhood for the set $\{P_{ij}\}$.

3.5   Give an example of a situation in which the application of a skeletonization algorithm is likely to be beneficial and one in which it is likely to have adverse effects.

3.6   A certain visual inspection process is to operate on images that contain two-dimensional shapes represented by binarized line segments of single pixel width. The process has the following requirements:

  (a) It must determine whether or not an object outline is connected.
  (b) It must be capable of measuring the Euclidean distance between any two points on the object profile.
  (c) It must be capable of applying a smoothing process to reduce noise effects.
  (d) It would be advantageous to have the capability to simulate the effect of horizontal/vertical shifts in the relative position of objects within the image.

  In the light of these considerations describe the factors that would determine the selection of an appropriate image array tesselation.

Table 3.1

Gray-scale level (g)	Number of pixels at level g
0	0
1	0
2	0
3	0
4	0
5	230
6	1020
7	1610
8	850
9	150
10	136
11	100
12	0
13	0
14	0
15	0

3.7 An imaging system generates images of 16 gray levels on a 64 × 64 pixel matrix for subsequent visual inspection and classification, and is interfaced to a small computer system provided with some purpose-written software for image storage, manipulation and display. Table 3.1 shows the pixel distribution across the available gray scale values for a typical image produced by the system. Explain why this image might be considered unsatisfactory.

Describe and evaluate the effects of a computer algorithm for contrast enhancement which would be likely to improve the visual appearance of this image and sketch its gray level histogram before and after transformation, commenting on the result.

4 ★ Structural operations on images

4.1 INTRODUCTION

We have seen how the appearance of patterns and images may be altered so that relevant (situation-specific) information may be extracted most easily. In almost all the earlier material, however, the common focus of attention in applying such techniques has been to deal with situations where the acquired and suitably transformed image has been intended for subsequent inspection by a human observer or where the image is to be subjected to automatic processing by machine.

The context of this book, however, was described as relating to the requirements of vision processing for robotic systems and other application areas involving machine/computer processing of patterns and pictures, and for the remainder of the text we shall be concentrating principally on ideas and techniques of primary applicability in this area.

More specifically, in trying to design an automatic system for interpretation or manipulation of images we are often concerned with operations that give us information about the *structure* of an object, and allow us to separate and distinguish different objects or areas of importance within an image, which allows us to discard information of no immediate relevance, and so on. In this chapter, therefore, we shall explore image operations that will help us to generate information about the content of images and to represent entities within images in a convenient way. As we shall see, in certain practical situations this will be all that is required to equip a vision-based robotic system with the means of carrying out its specified task, while, even in more difficult or wide-ranging tasks, such processing is a necessary preliminary to the application of more powerful interpretation processes.

4.2 IMAGES AND OBJECTS

In many applications of visual processing – in both human and computer terms as it happens – it is not necessary to deal with intricate or very subtle images (or, more precisely, such subtleties as are present are largely irrelevant to the task in hand). In such circumstances objects can often be represented by applying some process which dramatically reduces the

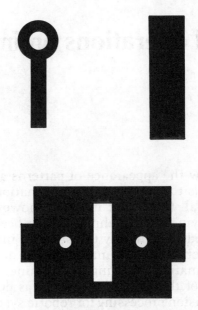

*Figure 4.1 Examples of easily distinguished binarized object
representations.*

amount of information it is required to handle. For example, we have
already indicated in a previous chapter some of the advantages of working
with binarized images where appropriate. To distinguish between the
mechanical components whose images are shown in Figure 4.1, for
example, binarized descriptions would be perfectly adequate.

Likewise, we realize intuitively from our everyday experiences that
many objects are identifiable purely from a very approximate description of
rough shape and relative dimensions. The examples shown in Figure 4.2
will serve to illustrate this point. Finally, to emphasize the idea of information reduction in object representation, it is useful to point out that in
many instances it is productive, particularly for more complex visual processing, to reduce the scale of a problem by breaking down an image into
sub-areas which can be dealt with individually, thereby, at least in principle, reducing a complex processing task to a sequential set of less complex sub-tasks. For example, in automatic document reading, it would
generally be more productive to work at the letter (basic unit) level than by
trying to recognize as a separate unit every word (or combination of basic
units) that could occur. Thus, THE would be processed as T → H → E, and
so on.

These observations lead us to the conclusion that it may be productive
in many applications of computer vision systems to work, where possible,
with general shapes of objects rather than the fine detail. More precisely,
this suggests that it is useful to be able to identify the *outline* of an object,

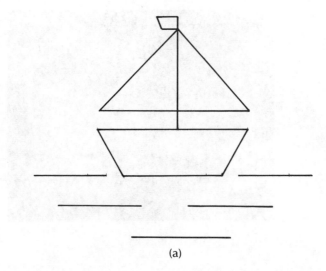

(a)

(b)

Figure 4.2 Representation of objects using minimal information: (a) Only straight-line segments make up this easily recognized object. (b) Cartoons typically use approximations of features to convey information.

from which basic shape descriptions and physical measurements might be derived. Consequently, we need to be able to carry out the sort of transformation shown in Figure 4.3, where a huge amount of information reduction has taken place while preserving the basic shape characteristics of the 'subject' of the image.

It is worth noting even at this stage that in purely practical terms such transformations immediately raise a subsidiary question. The transformed image of Figure 4.3(b), where the outline of an object is traced out in terms

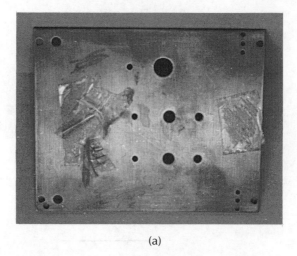

(a)

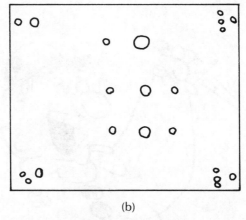

(b)

Figure 4.3 *Comparison of information in full gray-scale image (a) and its reduction to an object outline (b).*

of its boundaries, would be very inefficiently stored if its internal representation within a computer system were the same as that employed for the original gray-scale image. Of course, this is partly because the image is now binarized, but more specifically because by its very nature an image that contains only an object outline will be very sparsely populated in terms of black points. Consequently we shall in due course need to give further consideration to the question of the internal representation of such images.

In the following sections we shall be concentrating on these issues and the relationship between images as such, the objects within them, and the means of encoding the important information about objects such that an efficient representation is obtained.

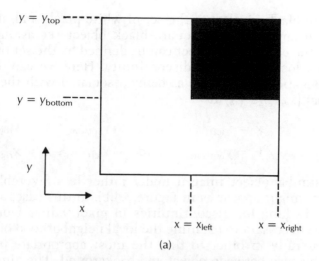

$y = y_{top}$

$y = y_{bottom}$

y

x

$x = x_{left}$ $x = x_{right}$

(a)

(b)

Figure 4.4 (a) Ideal and (b) more usually encountered 'edges'.

4.3 EDGE DETECTION

The derivation of the outline boundary of an object within an image is equivalent to identifying the *edge* that divides the object itself from its background. Algorithms that operate so as to identify the image pixels that lie on such an edge are known, not surprisingly, as *edge detection algorithms*, and examples abound in the research literature.

An intuitive grasp of how such edge detection algorithms might be formulated can be gained by considering the examples shown in Figure 4.4. In Figure 4.4(a) it is quite clear that there is a sharp discontinuity between the set of black pixels and the set of white pixels and, if the black pixels define an approximately square black object set against a white background, the edge of the object can be defined by the set of pixels that lie at the exact location of that discontinuity. Here we can define the precise spatial location of the edge as being associated with the set of pixels $\{S\}$ such that $(x_e, y_e) \in \{S\}$ if

$$x_e = x_{\text{left}} \qquad and \qquad y_{\text{bottom}} \leqslant y_e \leqslant y_{\text{top}}$$

$$y_e = y_{\text{bottom}} \qquad and \qquad x_{\text{left}} \;\; \leqslant x_e \leqslant x_{\text{right}}$$

A similar object imaged under rather less favorable circumstances, however, might appear as in Figure 4.4(b). In this case, although the principle of looking for discontinuities in pixel values would still apply, it would be necessary to examine the local neighborhood of individual pixels quite carefully in order to find the most appropriate place in which to locate the edge between object and background. The aim of standard edge detection algorithms, therefore, is to search for discontinuities in image intensity by weighing local evidence about the distribution of pixel values.

In practical terms, then, an edge detection operator could utilize a window which is moved across successive image points, at each point computing a value which, based on local neighborhood information, represents some measure of the probability that the point in question is, in fact, an edge point. An advantage of generating such a probability function is that we can then use this information in a variety of ways to decide exactly where the edge should be located. This will give valuable flexibility, since in real images it is not usually possible to be completely sure whether a discontinuity is caused by the presence of an edge or by noise-induced intensity changes, and inevitably some degree of uncertainty will prevail.

4.3.1 Edge detection operators

A common method of identifying the intensity discontinuities that mark out object edges involves computing a gradient function in respect of image intensity, on the basis that a high local intensity gradient, indicating a sudden intensity transition, is likely to be strong evidence for the existence of an edge discontinuity. To illustrate this, we shall first consider the *Roberts operator*, which involves a pixel-by-pixel computation based on the 2×2 window shown in Figure 4.5. For each possible position (x,y) of the window in the image a gradient function $G(x,y)$ is determined, given by

$$G(x,y) = \sqrt{[\{f(x,y) - f(x+1,y-1)\}^2 + \{f(x+1,y) - f(x,y-1)\}^2]}$$

As can be seen, the gradient measure $G(x,y)$ is based on the comparative intensities in orthogonal directions across the diagonals of the window,

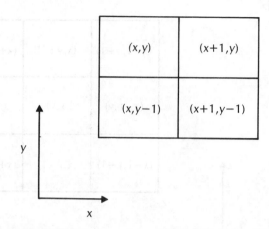

Figure 4.5 Definition of the processing window for the Roberts operator.

and can assume a value ranging from 0 (where all four pixels are of equal intensity) to a relatively large positive value depending on the intensity differences in the neighborhood of (x,y).

One obvious disadvantage of the Roberts operator, however, is that, principally because it is based on such a restricted pixel neighborhood, it can easily be affected by intensity changes caused by noise/random effects, leading ultimately to the identification of 'false' edges. One easy way of reducing the sensitivity of this type of operator to the effects of noise and hence help to avoid spuriously high gradient values is to attempt to take account of the intensity distribution over a rather larger area. In the context of the current approach this corresponds to considering a larger window over which to carry out the gradient calculation. Consequently, many other gradient operators could be considered, of which one common example is the *Sobel operator*.

The advantage of the Sobel operator is that it encompasses a degree of inherent neighborhood smoothing in addition to its primary function in evaluating a gradient value. As illustrated in Figure 4.6 the Sobel operator uses a 3×3 window computing a gradient function $G(x,y)$ given by

$$G(x,y) = \sqrt{[\{(f(x+1,y+1) + 2f(x+1,y) + f(x+1,y-1))}$$
$$- (f(x-1,y+1) + 2f(x-1,y) + f(x-1,y-1))\}^2$$
$$+ \{(f(x-1,y-1) + 2f(x,y-1) + f(x+1,y-1))}$$
$$- (f(x-1,y+1) + 2f(x,y+1) + f(x+1,y+1))\}^2]$$

It can be seen that this operator combines a gradient measure in the x-direction (the first { } bracketed term) with a similar measure along the y-axis (the second { } bracketed term). By varying the weighting factors and window size a whole range of other edge detection operators may be defined and, indeed, examples of such variations in the specification of edge operators abound in the literature.

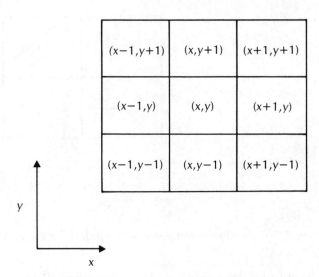

Figure 4.6 Definition of the processing window for the Sobel operator.

4.3.2 Edge identification

An edge detection operator will produce an array in which each point is assigned a value (*the edge value*) which approximates the probability that the point in question lies on an edge in the original image. The higher the edge value, the greater the probability that the point forms part of an object edge. In order to produce the edge picture itself (i.e. trace out the outline of the object of interest) it is necessary to decide which points in the edge array are to be considered as actual edge points and which are not. This requires the application of a now familiar thresholding procedure such that in relation to some operator-chosen 'edge threshold' value, θ:

$$G(x,y) > \theta \rightarrow \text{point } (x,y) \text{ is an edge point}$$

$$G(x,y) \leq \theta \rightarrow \text{point } (x,y) \text{ is not an edge point}$$

Working with 'ideal' images such as those presented in Figure 4.1 would, of course, allow an almost ideal edge picture to be generated by the selection of a low value of θ, since the gradient operator will be working with an image in which easily identified areas of uniform intensity are separated by sharp discontinuities, but in most practical cases the choice of θ must be such as to effect a compromise between ensuring that all real edges are identified, while not allowing false edges to appear because of noise, spurious contrast changes, and so on. Figure 4.7 shows a portion of an arbitrary gradient array (relating to an approximately circular object) and the effect of:

1. ideal choice for θ, generally unattainable in practice (Figure 4.7(b));

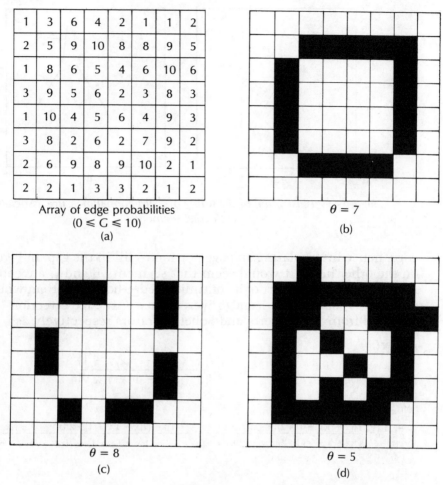

Figure 4.7 Relationship between edge probabilities and a thresholded edge picture.

2. choosing too high a value for θ (Figure 4.7(c));
3. choosing too low a value for θ (Figure 4.7(d)).

Again, in most practical situations, it is most likely that the outline produced by a gradient operator/thresholding sequence will be imperfect, containing both spurious edges and discontinuities in valid edges, and will consequently require further processing. This generally involves procedures based on smoothing and thinning operations, together with algorithms that identify localized edge-point groupings, to allow interpolation for gap-filling and so on. In this respect the 'direction' of an edge can provide useful information. For example, in Figure 4.8(a) a reasonable assumption might be that segments *A* and *B* form part of a continuous

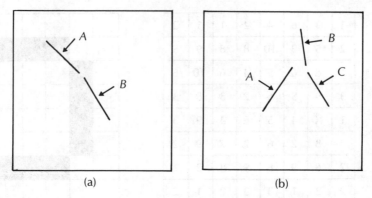

Figure 4.8 Illustration of the utility of a directional measure associated with edge segments.

edge, hence interpolating an edge portion to fill in the gap. In Figure 4.8(b), on the other hand, it would seem that segments *B* and *C* are considerably more likely to form part of a continuous edge boundary than would be the case if, say, segments *A* and *B* were joined. An edge direction (*α*) can be obtained from the Roberts and Sobel operators respectively as

Roberts

$$\alpha = \tan^{-1} \frac{f(x+1,y) - f(x,y-1)}{f(x+1,y-1) - f(x,y)}$$

Sobel

$$\alpha = \tan^{-1} \frac{\{f(x+1,y+1) + 2f(x+1,y) + f(x+1,y-1)\} - \{f(x-1,y+1) + 2f(x-1,y) + f(x-1,y-1)\}}{\{f(x-1,y-1) + 2f(x,y-1) + f(x+1,y-1)\} - \{f(x-1,y+1) + 2f(x,y+1) + f(x+1,y+1)\}}$$

4.4 REGION IDENTIFICATION

As we have seen, one way to identify object shape is to generate information about edges. Another way, which can be very effective in some applications, is to work with *silhouettes*. This is not fundamentally different from an edge description inasmuch as the assumption is made that overall shape characteristics, rather than fine detail or subtle shade variation, are of principal importance, as could be the case, for example, in identifying (or even determining gross characteristics such as surface area of) workpieces such as those shown in Figure 4.1. This method of reducing the information in an image to a shape representation of an object assumes that an object is defined by roughly uniform surfaces rather than by object/background discontinuities at edges, which amounts to much the same thing in practice.

It is fairly clear that this type of segmentation (the separation of specific objects or entities within an image) is easily accomplished by applying a thresholding procedure to the raw gray scale image. This was actually illustrated conveniently in the example given in the earlier discussion on binarization of images (see Figure 2.12). This procedure would, of course, work very well for situations where the images are simple (e.g. well defined dark objects of approximately uniform intensity on a light background), but could also give more detailed information about related regions of images in more complex cases. This is particularly true if other image characteristics apart from, or in addition to, intensity (perhaps texture, for example) are used in the region identification process. Indeed, one of the advantages of this approach is that in principle it can be applied to a variety of image attributes.

4.4.1 Regions and edges – contour tracing

There may be situations in which, although the preceding region identification technique is particularly applicable, there is still a need to generate edge information. For example, we shall see that an edge description of an object can lead to a very efficient means of representation for that object in a computer-based processing system. In such cases it is necessary to find a procedure that will generate an edge description from a silhouette or region-based description.

A convenient way to achieve this objective is to adopt some form of contour scanning or edge tracing algorithm. The principle of this approach is to find any point on the edge of an object and then, by utilizing local information, track around successive points on the edge (marking each point traversed) until the original starting point is reached. The trajectory of the tracker – the sequence of marked points – should then define the outline of the object. This requires an algorithm of the sort shown in Figure 4.9(a). Although straightforward in principle it has its drawbacks, particularly in terms of its ability to resolve rapidly changing edge directions and in its ability to cope with complex shapes where, for example, a generally uniform surface is interrupted by a hole or some other sub-region. Nevertheless, in appropriate circumstances this type of edge tracking can be very effective.

Figure 4.9(b) illustrates the principle involved in contour tracking, showing the approximated contour obtained from the application of the sort of algorithm specified.

4.5 GENERALIZED REGION SEGMENTATION

The previous example made the assumption that the identification of objects in terms of their basic shape could take place under constrained

Let {IN} represent set of pixels that define object
Let {OUT} represent set of pixels outside object

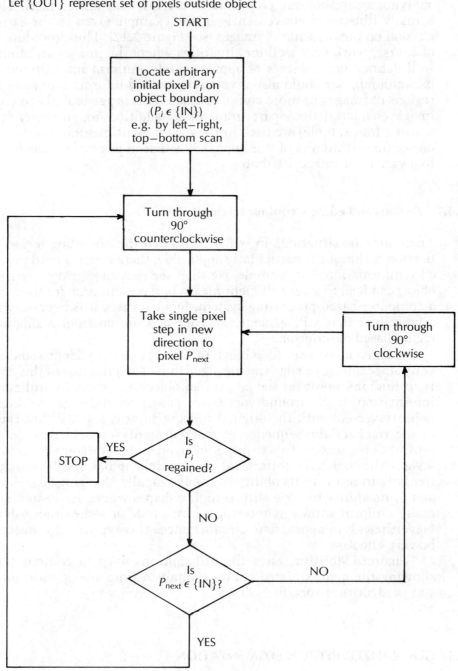

(a)

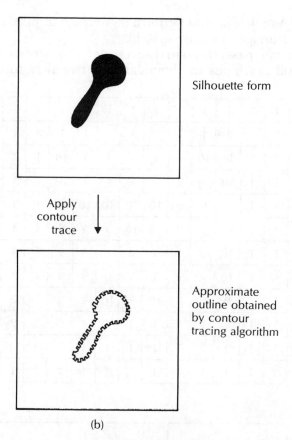

Silhouette form

Apply contour trace

Approximate outline obtained by contour tracing algorithm

(b)

***Figure 4.9** (a) Basis of a contour tracing algorithm and (b) the underlying ideas of its application.*

conditions (for example, objects have well defined and noncomplex struc-ture, and so on). In many cases, however, the segmentation of an image into distinct regions, within which pixel points may be considered as related in some way, requires a less simplified approach. For example, an image of a set of intricate electronic components might benefit from a segmentation process involving region identification beyond the capa-bilities of the procedure described above. Many techniques have been described to deal with these less constrained cases, but as a general rule the principles involved are based on the following approach, illustrated by reference to Figure 4.10.

Initially, a first pass scans the gray scale raw image and identifies primary regions by connecting together groups of pixel points of constant gray scale intensity value. (Here we are arbitrarily using intensity as the image attribute of interest though, as already indicated, others could be used.) In an illustrative small-scale example, this would take the raw image data of Figure 4.10(a) and perform a grouping of pixels to give the primary segmented image of Figure 4.10(b).

A second pass through the image now works with these primary regions and examines the boundaries between regions already identified.

1	1	1	1	1	2	1	1	1	1	1	1	1	1	1
1	1	1	1	1	2	1	1	1	1	1	1	1	1	1
1	1	10	10	1	2	1	1	1	10	1	1	1	1	1
1	1	10	10	1	1	1	1	1	10	1	1	1	1	1
1	1	9	9	9	10	10	10	10	10	1	1	1	1	1
1	1	9	9	9	10	10	9	9	9	1	1	1	1	1
1	1	10	9	9	4	6	6	9	9	1	1	1	1	1
1	1	10	9	9	5	5	6	9	9	1	1	1	1	1
1	1	9	9	9	10	9	10	9	9	1	2	7	7	1
1	1	1	1	1	1	1	1	1	2	2	2	7	7	1
1	1	2	2	1	1	1	1	1	2	2	2	1	1	1
1	1	1	1	1	1	1	1	1	1	1	1	1	1	1

(a)

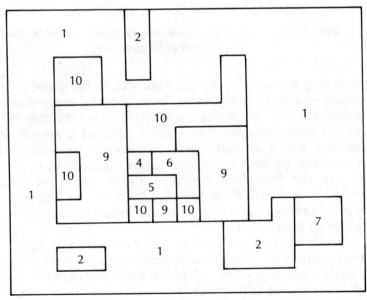

(b)

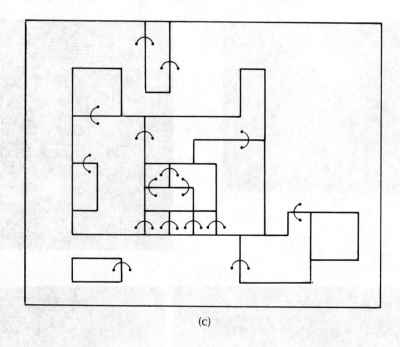

(c)

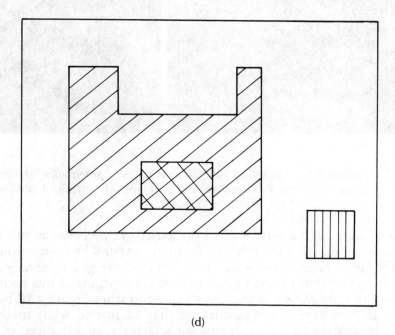

(d)

Figure 4.10 Progression of steps in a region segmentation process.

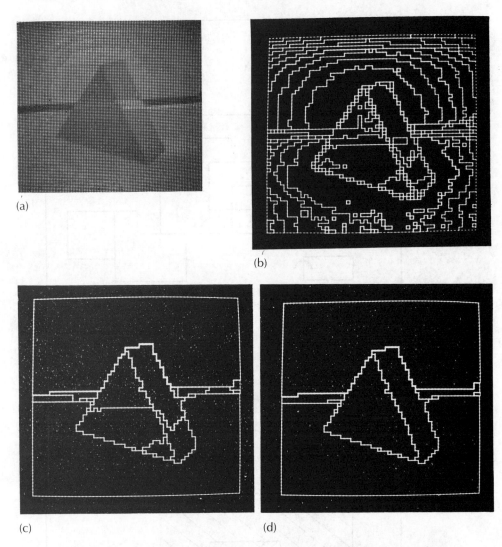

*Figure 4.11 A practical illustration of region segmentation. (From Brice,
C. and Fennema, C. 'Scene analysis using regions', Artificial Intelligence, 1,
1970, 205–226.)*

Depending on the differences in intensity value between primary regions
on either side of a boundary (a familiar thresholding operation will check
this) these primary regions may either be merged to give wider-ranging
regions or left as isolated pixel groups. An example of this second phase of
processing is shown in the transformation from Figure 4.10(b–c).

In this case the segmentation has identified what might easily be
interpreted as two components – a relatively large flanged object with a
slightly off-center region of a different shade (perhaps a depression or, more

likely in view of the relative intensity levels, a slightly raised area) together with a smaller square object (see Figure 4.10(d)). A more realistic illustration of the application of this type of process is given in Figure 4.11, showing an original gray scale image (a), the result of the primary region segmentation (b), and the segmented image obtained after region merging (c) and (d).

4.6 OBJECT REPRESENTATION AND OBJECT CODING

Attention has already been drawn to the fact that, when interest is focused primarily on the structure of an image in relation to object outlines, it is advisable to consider how best to encode and store a representation of that object efficiently. In particular, a conventional $N \times N$ pixel encoding (such as described in Chapter 3) of the image shown in Figure 4.12, say, would be extremely wasteful, since for the most part the representation would consist of strings of zero levels. Furthermore, successive manipulations of such encoded images, perhaps requiring a large amount of transferring of image data, is likely to incur unnecessarily time-consuming computational overheads. It is useful, therefore, to consider a more appropriate means of encoding the image data.

For cases such as this, a very useful technique for object representation is that of *chain encoding* the object of interest, a method originated by

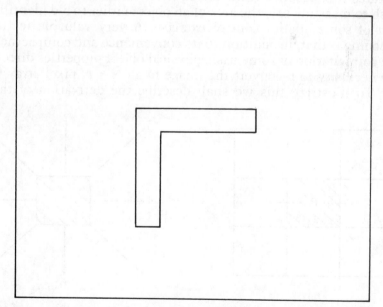

Figure 4.12 *Example of a sparsely populated image matrix where pixel-by-pixel storage might be relatively inefficient.*

H. Freeman. Chain encoding employs the principle of finding an arbitrary initial edge point which can, if required, be precisely located according to its spatial coordinates and subsequently storing, not the values of all pixels in the image and not even an exhaustive list of all remaining specific edge-point coordinates, but only the directional changes (quantized at the single pixel step level) required to follow the edge until the starting point is eventually regained. Thus, returning to a familiar idea, any pixel has eight defined neighbors of interest (Figure 4.13(a)), each neighbor located one 'step' away in one of eight directions specified by a numerical coding, labeled as 0, 1, 2, 3, 4, 5, 6 and 7 respectively (Figure 4.13(b)).

The outline of an object can then be chain encoded as a sequence of incremental directional transitions to trace around the edge of the object, conventionally, though not necessarily, in the clockwise direction. An example is shown in Figure 4.14, the object shown being represented by the chain code sequence

$$11000060666464444332$$

derived from Table 4.1.

It is clear that such an encoding scheme can significantly reduce the storage requirements for many images of practical interest in the present context. The chain-code technique, however, has an advantage beyond these immediate considerations. In a typical application of a vision system, for example, it will often be necessary to obtain measurements or identify features that describe some physical characteristics of an object. We may wish to measure, say, the area of an object either to aid identification or as part of some quality control exercise. A very valuable feature of chain encoding is that, in addition to its convenience and compactness, it allows the computation of some basic physical object properties directly, without the necessity to reconvert the image to an $N \times N$ pixel array format.

To illustrate this we shall describe the derivation of the following

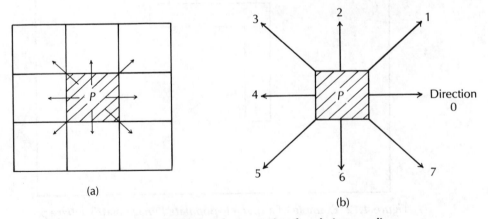

(a)

(b)

Figure 4.13 Directional convention for chain encoding.

Table 4.1

	Step sequence	Encoded direction	
START →	1	↗	
	1	↗	
	0	→	
	0	→	
	0	→	
	0	→	
	6	↓	
	0	→	
	6	↓	
	6	↓	
	6	↓	
	4	←	
	6	↓	
	4	←	
	4	←	
	4	←	
	4	←	
	3	↖	
	3	↖	
	2	↑	Start point regained

three physical parameters which may be useful in describing the properties of an object: object *perimeter*, object *width* (measured here as maximum width in the horizontal direction), and object *area*, in each case referring to the specific object shown in Figure 4.14 by way of example.

4.6.1 Calculation of perimeter

It is clear that an incremental step in any of the directions 0, 2, 4, 6 of the chain code corresponds to moving by exactly one pixel distance to give a contribution of 1 unit of distance in a perimeter measurement. At any point in the chain code defined by a transition in one of the directions 1, 3, 5 or 7, however, the corresponding distance traveled as $\sqrt{2}$ units (see Figure 4.15).

Consequently, it is possible to calculate the perimeter of a chain-encoded object by stepping through the elements of the chain code and accumulating the sum of the total number of steps of *even* direction (S_E) and the number of code elements of *odd* directional value (S_O). The object perimeter, P, is then given by the computation of

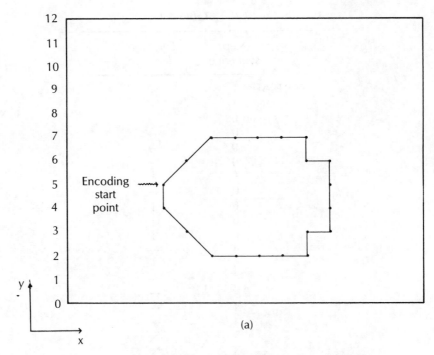

Figure 4.14 An example of the chain encoding procedure.

$$P = S_E + \sqrt{2}S_O \quad \text{units}$$

This can be directly related to a quantitative physical measurement in practical units provided that the characteristics of the imaging device are known.

In the example shown in Figure 4.14, this procedure generates:

$$S_E = 16$$

$$S_O = 4$$

Hence,

$$P = 16 + 4\sqrt{2}$$

$$= 21.66 \quad \text{units}$$

This measurement can be checked directly on Figure 4.14 using the coordinate scale shown.

4.6.2 Area measurement

Object area is obtained by direct computation from the chain code by evaluating the elemental contributions of vertical strips bounded by the

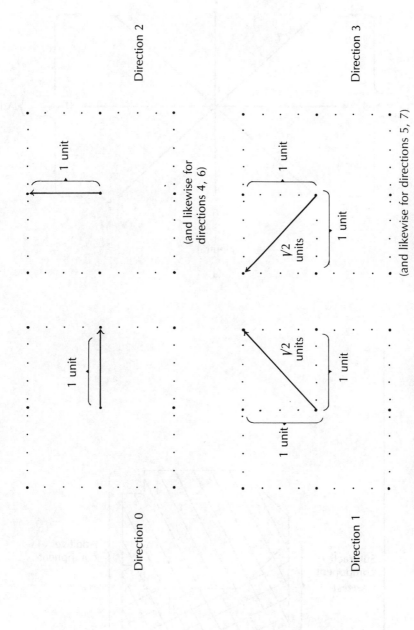

Figure 4.15 Relationship between chain element direction and transition distance.

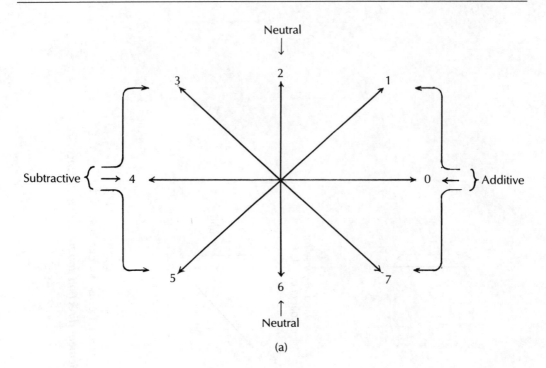

(a)

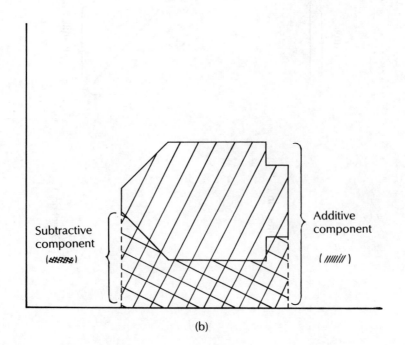

(b)

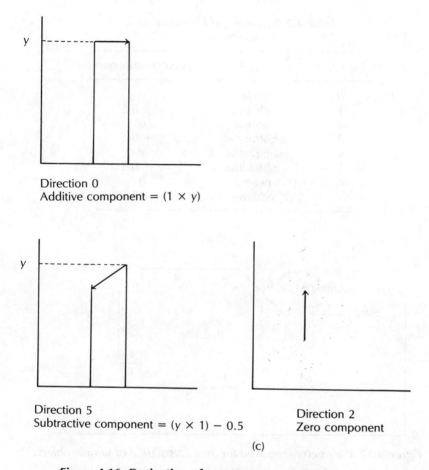

Direction 0
Additive component = (1 × y)

Direction 5
Subtractive component = (y × 1) − 0.5

Direction 2
Zero component

(c)

Figure 4.16 Derivation of component contributions of chain elements to object area evaluation.

current directional transition path between two successive edge pixels and the x-axis. Increments are additive, subtractive or of zero contribution according to the notation of Figure 4.16(a). The method corresponds to a step-by-step execution of a procedure shown in Figure 4.16(b) which calculates the object area by subtracting the sub-area identified by double cross-hatching from the area identified by the single cross-hatching. For example (see Figure 4.16(c)):

1. A step in the 0 direction adds an elemental area of (1 × y) units to the overall area calculation.
2. A step in the 5 direction would correspond to a *subtractive* component of (y − 0.5) units in the area calculation.
3. A step in direction 2, which follows a vertical path, contributes a zero component to the area calculation.

**Table 4.2 Summary of elemental area
contributions**

Direction	Type	Area contribution
0	additive	y
1	additive	$y + 0·5$
2	neutral	0
3	subtractive	$y + 0·5$
4	subtractive	y
5	subtractive	$y - 0·5$
6	neutral	0
7	additive	$y - 0·5$

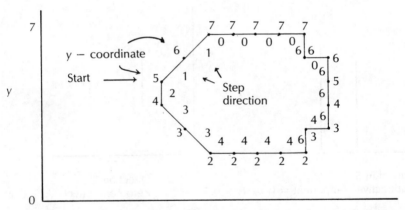

Figure 4.17 Parameters required for area calculation of sample object.

Applying these principles to the complete set of possible directions generates the table of contributions shown in Table 4.2, and taking the specific example redrawn in Figure 4.17 yields the calculation of this object's area (A) as:

$$A = 5.5 + 6.5 + 7 + 7 + 7 + 7 + 0 + 6 + 0 + 0 + 0$$
$$- 3 + 0 - 2 - 2 - 2 - 2 - 2.5 - 3.5 + 0$$
$$= 29 \text{ square units}$$

Again this can be checked by direct reference to the diagram.

4.6.3 Object width

The maximum width of an object measured across the horizontal can be obtained in a similar manner by observing the overall distance traveled in incremental steps in the x-direction as the chain code is traversed. If the horizontal distance is cumulatively generated at successive incremental

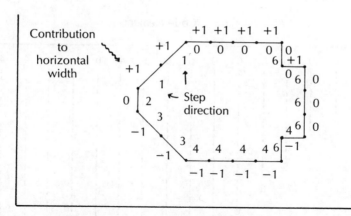

Figure 4.18 Parameters required for width calculation of sample object.

steps, a measure is obtained of the position of the extreme edges of the object.

In terms purely of horizontal distance, the directions 0, 1 and 7 each correspond to a 1 unit shift to the right (+1), while each step in directions 3, 4 and 5 corresponds to a 1 unit shift to the left (−1), and directions 2 and 6 are orthogonal to the x-axis and give no x-shift (0 units).

To determine the x-width, therefore, the calculation proceeds as follows:

1. Locate the starting element of the chain code.
2. Calculate the x-shift associated with the first chain element traversal.
3. Calculate the x-shift associated with the first *and* second chain element traversals.
4. Calculate the x-shift associated with the first, second and third chain element traversals; and so on for all elements in the chain code.

Let the maximum horizontal shift at any step $= S_{high}$.
Let the minimum horizontal shift at any step $= S_{low}$.

Then the horizontal width of interest (W) is given by

$$W = S_{high} - S_{low}$$

Applying this procedure to the example of Figure 4.18 gives the cumulative table shown in Figure 4.19, and hence:

$$S_{high} = 7$$

$$S_{low} = 0$$

$$W = 7 - 0$$

$$= 7 \text{ units}$$

Again this is verifiable as correct by direct comparison with the diagram.

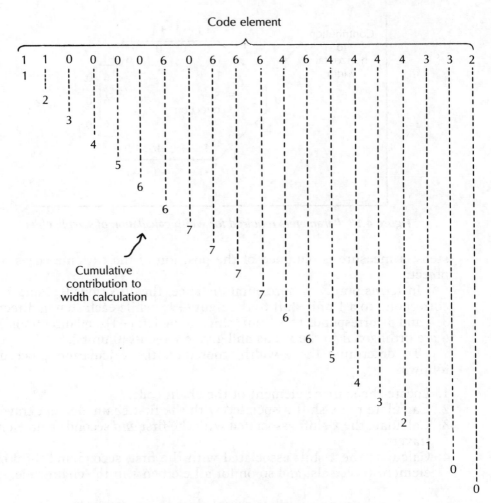

Figure 4.19 Execution of width calculation algorithm.

4.7 STRUCTURAL CHARACTERISTICS AND THE HOUGH TRANSFORM

Of course, in trying to compute information about the structure of objects of interest within images in an attempt to determine image contents, other possible means may be considered and, in particular, we may wish to focus on structural features in a more qualitative, rather than explicitly quantitative, way.

For example, we may wish to identify the occurrence of straight line segments in an image, since this might be helpful in a number of ways. It could allow us to approximate the shape of some object as a piecewise linear reconstruction of its outline, for example, or it might help us in the task of operating on a nonideal edge picture to improve the appearance and accuracy of the processed image. It would be particularly useful if we could in some way try to weigh the evidence available in considering whether or not a number of image points should be regarded as individual elemental components that contribute to a straight line feature, perhaps allowing us to interpolate points that are missing as a result of imperfections in some earlier processing algorithm. So, either to improve the quality of representation of imaged objects or to extract feature information about image content, a means of identifying basic shape information such as straight line segments could be very valuable. An approach that has been successfully used in this type of application is the Hough transform which we shall now examine. Although the technique may be extended for the computation of shape information of greater complexity, we shall consider its operation here only in the detection of straight line segments.

The technique is based on the interrelation between points in two-dimensional space represented as their (x,y) spatial coordinates (which we shall call xy-space) and their representation in a transformed 'straight line space' (which we shall call mc-space). Any point in xy-space that is to be considered as part of a straight line segment must, of course, satisfy the general equation

$$y = mx + c$$

and this must be true for all points on a given straight line.

By rearranging this general equation it is easily seen that a given point of interest must also satisfy the equation expressed as

$$m = y/x - c/x$$

and thus the point could be thought of as being transformed into a straight line in the space with axes formed by m and c. This is the mc-space introduced above.

Thus any point in xy-space (see, for example, Figure 4.20) can be mapped to a line in mc-space (see the corresponding representation in Figure 4.21). But, of course, a set of points that lie on a straight line segment in xy-space must each individually satisfy the $y = mx + c$ equation and will therefore generate in the transformed mc-space lines that intersect at a common point. Consequently, if we take an arbitrary set of points in xy-space, an examination of their transformed equivalent in mc-space will allow us to determine, just by examining the number of common intersection points, the likelihood that groups of points lie in straight line segments. Furthermore, it may be assumed that, at a given point of intersection in mc-space, the greater the number of lines intersecting, the greater the weight of evidence that a straight line segment exists in xy-space.

It is necessary, therefore, only to keep track of the intersection points in the transformed space to compute a measure of 'line-segment probability' in the original image. This can be achieved by defining an array with

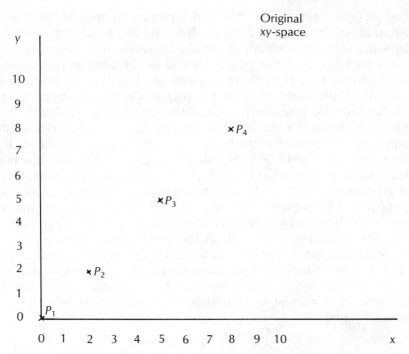

Figure 4.20 Representation of xy-space with sample points for illustration. $P = (x,y)$, $P_1 = (0,0)$, $P_2 = (2,2)$, $P_3 = (5,5)$, $P_4 = (8,8)$.

elements corresponding to points spatially distributed over the space defined by the m,c axes, and incrementing an array element for every crossing of a given spatial location by a line in mc-space. From this, in much the same way as we achieved the 'edge-probability' to 'edge-picture' mapping described earlier, a straightforward thresholding procedure will permit the detection of straight line features occurring in the original image.

Adding this type of technique to the techniques already discussed, we now have at our disposal a basic set of tools that relate to the representation of image objects, the detection of features of interest, and the computation of some quantitative descriptive features, all of which will be needed if we are to move on to approach the questions of identifying and interpreting images.

4.8 SUMMARY

This chapter has moved the emphasis of the discussion from direct questions about the appearance or quality of an image to an increasing

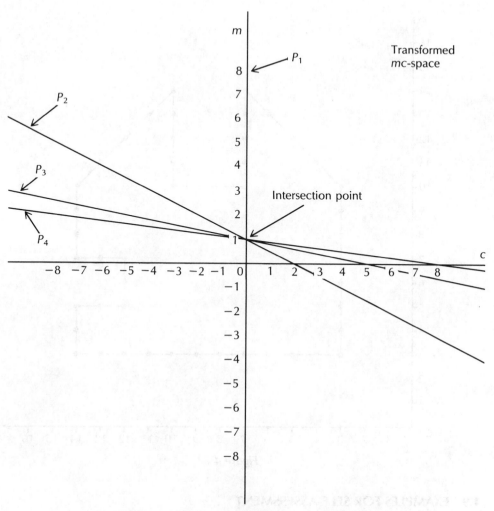

*Figure 4.21 Transformed mc-space for sample points shown in
Figure 4.20.*

concern about its actual content and a consideration of how to extract information appropriate to its interpretation.

The techniques described are likely to provide the means of obtaining information relevant to the recognition of an object or the identification of some specific characteristic of interest. Exactly how such information might be utilized will form the basis of later parts of this book. Before moving into that area, however, we shall briefly consider in the next chapter some important fundamental questions concerning the structure of processing hardware for carrying out basic image processing operations.

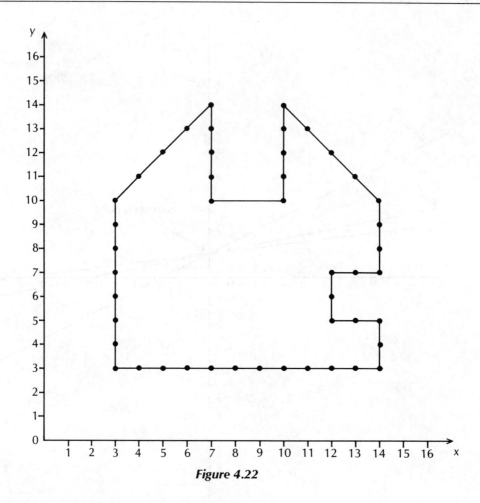

Figure 4.22

4.9 EXAMPLES FOR SELF-ASSESSMENT

4.1 For a number of sample images, investigate in more detail the way in which the transformation from an array of edge values to a thresholded edge picture affects the representation of objects in an edge-processed image.

4.2 Determine and describe a possible way in which different edge detection operators might be evaluated on a comparative basis, indicating particularly how such a comparison might be carried out in a quantitative way.

4.3 Derive a chain-coded representation of the image shown in Figure 4.22.

Explain how this chain code could be modified (without recourse to a complete reconstruction of the original image) so that the object outline can be traversed in the opposite (i.e. counterclockwise) direction.

4.4 Starting from its chain-coded representation, calculate the area of the object encoded in Question 4.3.

4.5 Explain, in the context of machine vision, the advantages of representing objects in terms of information about their outline shape, briefy explaining how such a representation may be generated.

4.6 Consider an industrial inspection process in which an imaging system, which generates an outline representation (single pixel width) of objects on a 128 × 128 pixel array, is working with component parts which are known to be approximately circular in shape but of varying sizes.

Discuss (quantitatively where appropriate) the relative merits of storing the image data obtained:

(a) directly as a pixel array,
(b) as a chain-encoded object representation.

Describe, stating any assumptions made, an algorithm for each case which would allow object area to be estimated.

4.7 Describe the potential practical difficulties associated with the application of the Hough transform in the form described in the text.

How might these difficulties be overcome?

5 ★ Computational architectures and image processing

5.1 INTRODUCTION

Chapters 2–4 have discussed some common operations applied to digital patterns and pictures and described typical algorithms to carry them out. In practice, of course, it is necessary to implement these algorithms, and it is important to make the point that an implementation could be embodied in a variety of forms. At one end of the spectrum, it is commonly assumed that most processing will be carried out by software running on a conventional computer system. At the other extreme, perhaps, it might be possible to construct an entirely self-contained and specially designed piece of hardware to execute some process of interest. In between these two approaches, it is possible to find many hybrid systems which combine hardware and software elements to give the best of both options in a particular situation. Indeed, the implementation of an algorithm as a program running on a conventional computer itself constitutes a mixed hardware/software solution, and it is inevitable with the present state of technology that this option will be very common. The fact remains, however, that software requires hardware for its execution and in all the above options, therefore, the structure of the underlying hardware will be of fundamental importance in determining the efficiency and effectiveness of implementation.

In this chapter we shall address some basic questions concerning computational architectures that are appropriate for the implementation of image processing and related algorithms. Naturally we shall begin with the conventional serial digital computer, but our discussion will move on to consider alternative architectures which might be better suited to problems in this area. To identify ideas of importance and features of relevance we shall examine three computational structures – the serial computer, a cellular parallel processing structure with an array architecture, and finally a much more all-embracing theoretical computational scheme.

5.2 IMAGE PROCESSING WITH A SERIAL COMPUTER

Serial processing on a digital computer constitutes probably the most common medium for the implementation of pattern/picture processing

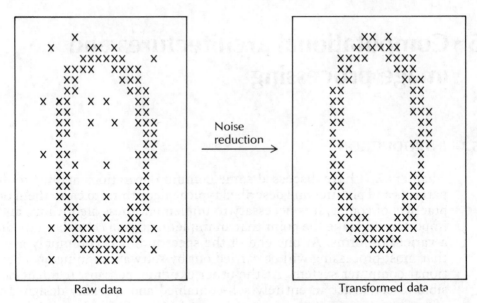

Figure 5.1 An example of the application of a noise reduction algorithm.

operations. It is clear that there is much to commend this approach, since such hardware is readily available and generally cheap and, provided that care is taken over the choice of programming language and the adoption of an appropriate programming and data organization methodology, software implementation is likewise readily achieved. While this cannot be denied, it is nevertheless the case that a serial processing approach is not necessarily optimal for many typical image processing tasks, because of inherent limitations in the structure of the underlying architecture.

This point is best illustrated by an example. Let us consider the implementation of an algorithm intended to remove isolated black points from a binarized image on a two-dimensional pixel array. This requires that all pixels are examined in relation to their immediate neighbors and, applying appropriate connectivity criteria, any black pixel that is found to have no black neighbors connected to it is changed to a white pixel. The result of such an operation would be to effect the transformation shown in Figure 5.1. If we assume that the array is of square tesselation and that the 8-C rule of neighbor connectivity is adopted, an algorithm such as that shown in Figure 5.2 is required for the processing of an arbitrary pixel. Now, of course, the computational complexity of this processing operation will depend on a number of factors – for example, how the pixel data is organized and stored in memory – but it is clear that, since

1. the procedure shown must be applied *to each pixel in turn* ($N \times N$ times in other words, remembering to make special arrangements for dealing with edge pixels),

2. in general, a number of similar operations are likely to be carried out on an image,
3. many different images may require such extensive processing in a practical task,

a very large number of computer operations are unavoidable, just because of the inherent nature of pixel-by-pixel processing and the serial limitation of the conventional digital computer.

We can observe, therefore, that although the serial machine has many advantages arising from the fact that it is general purpose, of relatively low cost and so on, the inherently iterative nature of many image processing operations suggests that it may not necessarily provide the most efficient computational structure for this type of task, particularly when time constraints are important. We may further note that this type of structure is certainly not like that which nature has provided for biological systems

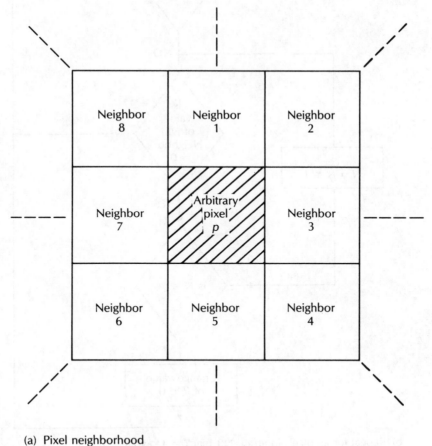

(a) Pixel neighborhood

Figure 5.2 Basis of the formulation of a noise reduction algorithm.

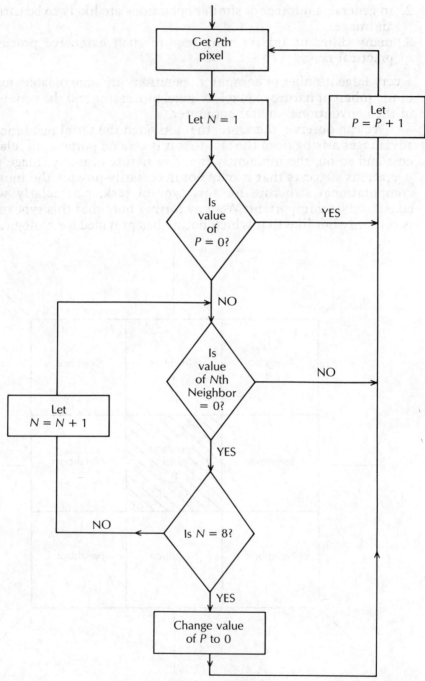

(b) Repeat for all pixels in array. NB black = 1, white = 0

Figure 5.2 Continued

and that these are, after all, precisely the sort of structures we are endeavoring to imitate at a functional level.

5.3 ALTERNATIVE COMPUTATIONAL STRUCTURES FOR IMAGE PROCESSING

We have seen that the serial processor is inherently limited in the way in which its internal structure matches the 'problem structure' of many image processing tasks. On the other hand, it should already be apparent, both from even a rudimentary knowledge of biological vision processors and from the immediately preceding discussion, that a degree of parallelism in the computational architecture could provide a much more effective correspondence with the inbuilt parallelism in many tasks of interest. In the present context, however, it is convenient to formalize a descriptive framework within which to discuss and identify the characteristics of potentially suitable architectures. Just such a classification scheme has been identified by looking, not specifically at the detailed logic-level description of computational hardware, but rather at the relationship between instruction execution and flow of data segments. At this level it is possible to identify three structural categories which are of particular relevance to the present discussion.

The conventional digital computer with its so-called serial architecture is then seen as a structure that operates on a stream of data taking one 'chunk' at a time, executing single instructions sequentially on successive data segments. This is illustrated diagrammatically in Figure 5.3(a), and in the adopted classification scheme such a structure would be labeled as a single instruction, single data stream (SISD) architecture.

A completely different architecture such as that illustrated in Figure 5.3(b) might, however, be envisaged. Here the computation is carried out by a group of processors, each operating on a different segment of data, and each executing its own localized instructions (although the processors are expected to communicate with each other in some way). This is, of course, a much more complex structure than the SISD case and is identified as a multiple instruction, multiple data stream configuration (MIMD) architecture in the classification scheme.

Somewhere in between the degrees of complexity implied by the two preceding examples it is possible to identify a third alternative. This corresponds to the single instruction, multiple data stream (SIMD) structure, and is illustrated in Figure 5.3(c). As can be seen, the basic principle involved here is that once again processing is carried out in parallel by a group of (interconnected) processors, each operating on an individual segment of data, but in this case each processor is made to execute simultaneously an identical instruction. The idea of carrying out an identical operation on a number of individual data chunks is familiar from the

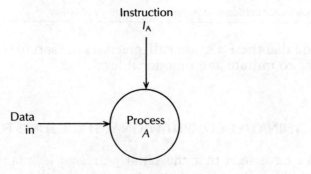

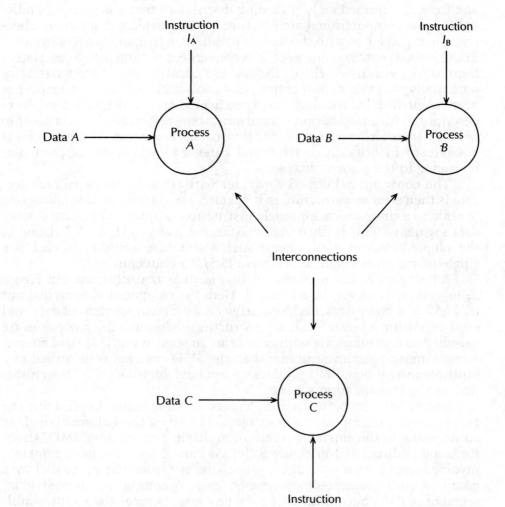

(a) SISD

(b) MIMD

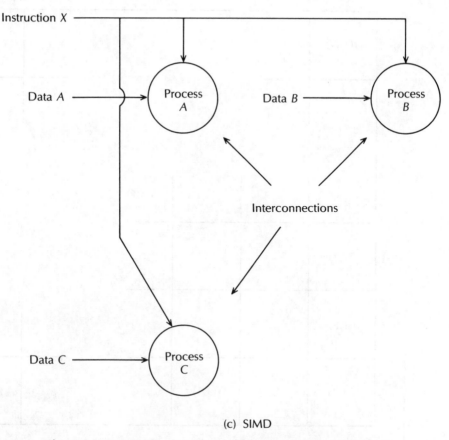

(c) SIMD

Figure 5.3 Categorization of possible processing architectures.

discussion of the noise removal algorithm considered earlier and, indeed, this type of activity is very common in many image processing tasks.

It is easy to see why SIMD architectures offer processing structures that are particularly well suited to image processing applications and why a number of practical systems based on these principles are now beginning to emerge as viable alternatives to serial machines. In the next section we shall examine such architectures in a little more detail.

5.4 *SIMD* ARCHITECTURES FOR IMAGE PROCESSING

A typical SIMD processor is designed to operate on an array of pixel points such as that shown in Figure 5.4. Here, an arbitrary point is identified, together with its immediate neighbors (assuming an 8-C connectivity rule). The basic philosophy underlying the SIMD architecture is that each

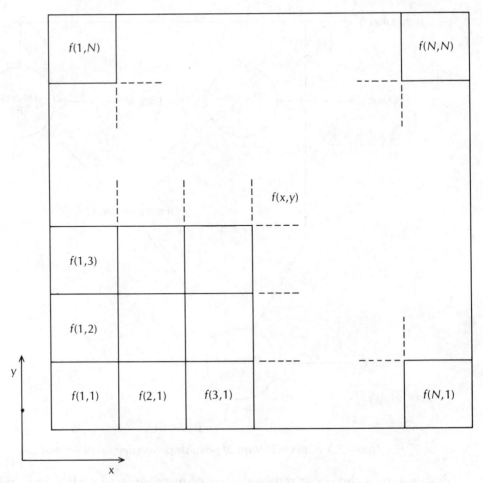

Figure 5.4 Image matrix with reference pixel labeling.

pixel in the image array has, in the actual processing device, its own individual associated processing element or cell, as illustrated in Figure 5.5, the overall system thereby comprising an array of processing cells which has a direct correspondence with the image array on which it is intended to operate.

Each cell is able to communicate with a set of neighboring cells (usually either its 4-C or its 8-C neighbors) so that it receives information about the value of its associated pixel point and its immediate local neighborhood, and is likewise able to propagate information about its own pixel value back to its defined neighbors. Furthermore, bearing in mind the structural principles of the SIMD configuration, each cell in the array is simultaneously instructed to execute an identical operation, thereby allowing

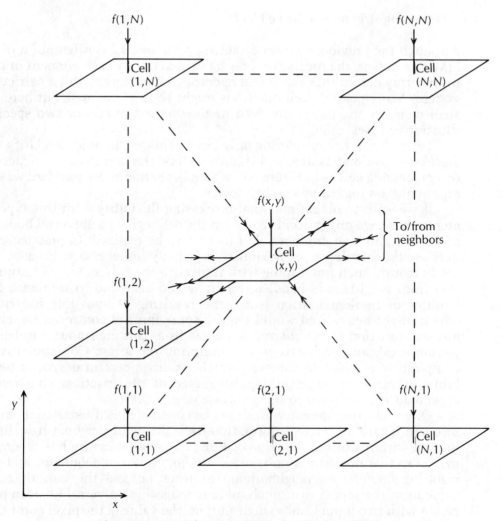

Figure 5.5 Spatial arrangement of cells in an SIMD array.

the parallel execution of a chosen operation simultaneously at every pixel point in an input image.

This type of architecture is clearly structurally well suited to the sort of typical image processing algorithm we have previously been discussing. It is also useful to point out a further advantage of this type of architecture which is that its inherently regular design makes it a particularly suitable candidate for implementation using VLSI technology, and, for this reason also, SIMD processing arrays are of great interest in image and pattern processing.

5.4.1   Processing structure at the cell level

Although the previous section established the overall configuration of an SIMD processor, the precise processing carried out by each element or cell in the array determines the actual operating characteristics of a particular system. Many possible cell functions might be implemented, but here we shall consider the issues involved in the context of one or two specific illustrative cases.

Let us begin by considering only binary images. In order to clarify the discussion, we may isolate and slightly redraw the input/output structure of a processing cell as in Figure 5.6, which is a rather more standard way of representing a logical processing device.

If we wish to obtain maximal processing flexibility with this type of architecture, we might decide to design the cell logic as a universal Boolean processor, allowing *any* possible function to be realized. In practice, not only would this degree of flexibility be largely unnecessary, because we rarely require such fine tuning with respect to the way in which information from neighbors is handled, but it would also lead to immense difficulties of implementation (since the resulting 2^{2^9} possible functions which might be realized would require some form of controller to select just one function at any instruction cycle from this enormous number of possibilities), and problems in programming the array. Consequently, a compromise is generally adopted which retains a certain degree of flexibility in function while taking advantage of the practical constraints expected to be involved in image processing tasks.

One such compromise which may be considered is illustrated in broad terms in Figure 5.7. Here, information from the eight neighboring cells is merged within a thresholding procedure (T is a variable which is programmable) so that no distinction is made among the set of neighbors, and the resulting compressed neighborhood information and the current pixel value form the inputs to a much more manageable universal Boolean processor with two inputs and a single output (the value of the pixel point that will be assumed at the next operational step).

It is easy to see how this cell logic can, by judicious selection of the variable parameters (through a suitable programming procedure) applied across the whole array, execute fairly powerful processing operations.

For example, selecting

$$T = 1$$

(i.e. $X = 1$ if any neighboring point $= 1$) and

$$\text{Boolean function} = \text{AND}$$

(i.e. new value of pixel $= 1$ if and only if current value $= 1$ *and* $X = 1$, indicating that the value of at least one neighbor $= 1$) will implement directly the noise reduction algorithm previously outlined. The point of

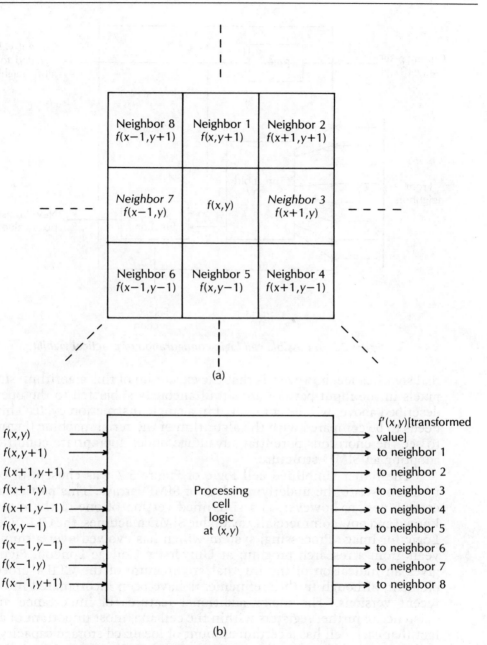

Figure 5.6 A maximally flexible input/output structure for processing cell design.

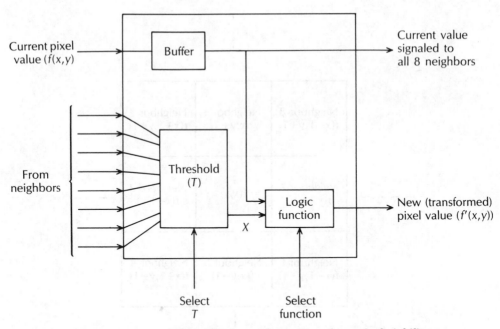

Figure 5.7 A possible cell logic configuration of practical viability.

real significance, however, is that the execution of this algorithm, since all pixels in the input pattern are simultaneously subjected to the operation described above, will be completed in a single instruction cycle. This only needs to be compared with the algorithm of the serial machine (Figure 5.2) to see the enormous potential advantage under appropriate conditions of adopting an SIMD structure.

The rather simplified cell logic of Figure 5.7 was chosen, of course, to demonstrate the underlying ideas of SIMD arrays. This particular processing design, however, is a simplified version of one of the most well known and now commercially available SIMD machines, the CLIP (Cellular Logic for Image Processing) system which has evolved over a number of years from a research program at University College London. Figure 5.8 gives an indication of the internal architecture of the CLIP4 processing element, although further refinements have been incorporated into more recent versions. The major additional features of importance are the existence of further registers within the cell and, most important of all, the fact that each cell has a certain amount of localized storage capacity. This makes it possible to handle gray scale images very conveniently (by considering *planes* of image data) and also means that straight numerical operations can be implemented relatively easily.

Although the CLIP system is perhaps one of the best known processors of its type it is not the only one. For example, the ICL DAP array processor has the useful feature that it can be programmed in a fairly conventional

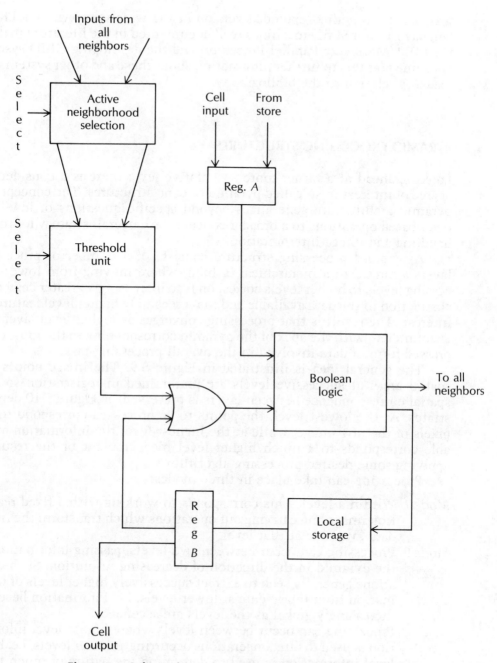

Figure 5.8 Schematic representation of the cell logic of the CLIP processor.

way using a modified/extended version of the widely known FORTRAN language. Other SIMD machines well documented in the literature include the MPP (Massively Parallel Processor) and the developing GRID system. For some pointers to further information about these and other systems the reader is referred to the bibliography.

5.5 PYRAMID PROCESSING STRUCTURES

Looking ahead at a rather more speculative level there is a considerable degree of interest in so-called pyramid or cone structures. The concept of a pyramid architecture generalizes beyond specific questions of low-level pixel-based operations to a broader concern with overall strategy for image handling and image interpretation.

A pyramid processing structure consists of a number of processing layers arranged in a hierarchical fashion, where moving from lower processing levels to higher levels corresponds to increasingly greater degrees of abstraction in the data available and in increasingly higher level features of interest. The result is that processing converges as higher level layers are encountered, with the apex of the pyramid corresponding to the most compressed form of data involved in the overall processing task.

The general idea is illustrated in Figure 5.9. The image points and spatial areas at successive levels are maintained in registration so that spatial correspondence between layers is preserved, as Figure 5.10 demonstrates. At the lowest level, the points to be processed correspond to the pixels of the raw image, while at the highest level the information available corresponds to a much higher level measurement or the result of applying some desired processing algorithm.

Processing can take place in three modes:

Mode 1 Within a level. This corresponds to working with a fixed resolution image and carrying out operations which transform the image data available at that level.

Mode 2 Processing can occur between two levels, passing information *up* the pyramid in the direction of decreasing resolution. Such operations generally seek to extract successively higher levels of information from image data at lower levels, i.e. information becomes increasingly global as the levels are ascended.

Mode 3 Processing can occur between levels where higher level information is used to direct operations occurring at lower levels, i.e. high-level information is used to determine the nature of much more local processing.

An idea of the basis of this architectural scheme can be obtained from a brief consideration of an envisaged application. Almost any of the standard

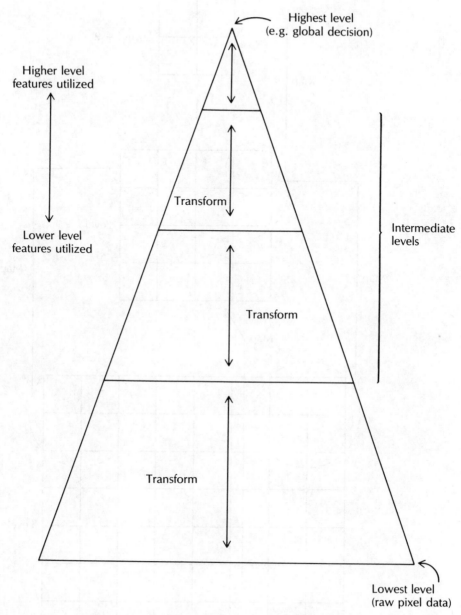

Highest level
(e.g. global decision)

Higher level
features utilized

Lower level
features utilized

Intermediate
levels

Transform

Transform

Transform

Lowest level
(raw pixel data)

Figure 5.9 Basic structure of a pyramid processor.

windowing operations (e.g. edge detection) could be described as a Mode 1 operation. In Mode 2 successive reductions of resolution where, say, a smoothing function maps from a group of gray scale points at level l to a single averaged value at level $l + 1$, could produce a series of data planes with increasingly high-level features, where local details are largely

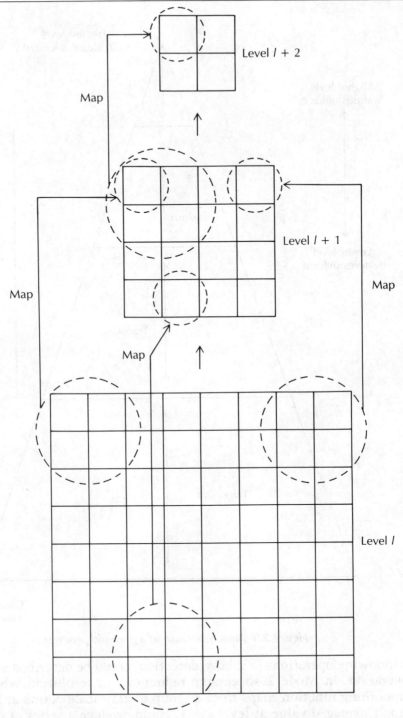

Figure 5.10 Interaction between processing levels in a pyramid structure.

suppressed. Mode 3 operation could allow a thresholding value to be obtained efficiently either by considering only the highest level of the pyramid to provide a global thresholding value, or by considering intermediate areas represented at lower levels within the pyramid which could be used for localized thresholding values at the lowest or raw data level.

The idea of a processing pyramid is principally conceptual, though a well developed theory of such an approach is now beginning to emerge. A major reason for including an overview of the conceptual basis for the approach is to demonstrate the methodology and indicate particularly how it might be possible to consider different specific computational architectures at different levels which could best match the processing tasks required.

5.6 SUMMARY

Whatever view is taken of the question of choosing an appropriate implementation for an image processing system, it is apparent that significant processing power and computational resources will be required for most practical applications. Until fairly recently effort has largely been concentrated on maximizing software efficiency in realizing algorithms in a form suitable for execution on a conventional serial computer. This chapter has attempted to show some of the alternatives and to indicate directions in which current efforts are moving. The bibliography contains further detailed pointers to the growing literature on computational architectures for image processing and related applications.

5.7 EXAMPLES FOR SELF-ASSESSMENT

5.1 In the context of pattern processing discuss the principal practical difficulties associated with the following:

(a) the implementation of a pattern smoothing algorithm with an SISD processing architecture,

(b) the use of an SIMD processing architecture as a general-purpose processing system.

5.2 For an SIMD logic array to process binarized two-dimensional images, discuss:

(a) the relative merits of using a 4-C or an 8-C connectivity rule in specifying the interconnection of constituent cells,

(b) the main design criteria that might be used to determine the detailed structure of the hardware.

Explain how the designer of such a system might reach a compromise between hardware complexity and functional generality in the design of the cell logic.

5.3 Stating any assumptions made, obtain an estimate of the relative processing times for the execution of the noise reduction algorithm described in the text for an SISD and an SIMD processing structure respectively.

6 ★ An introduction to pattern recognition

6.1 INTRODUCTION

We now move into the second main section of the book which deals more explicitly with questions about interpretation of images. Of course, some of the ideas introduced in earlier chapters have a significant bearing on the present discussion. In a simple situation – for example, in making a decision as to which of just two possible shapes an image represents – a straightforward measurement such as the computation of object area from its chain-coded representation might conceivably be equivalent to interpretation provided that the shapes are 'cooperative' in terms of their surface areas. At the opposite extreme, in a situation where an image represents a much more complex visual scene – perhaps in an aerial photograph – then a series of operations and measurements involving a very large amount of computational effort might be required in order to determine what the image contains and to 'interpret' it in the broadest sense.

However, many practical problems fall between these two extremes. Examples that readily spring to mind would include such areas as quality control in an automated manufacturing process (automated visual inspection of workpieces), identification of a restricted range of components on a conveyor belt, reading of labels in an automated warehouse, and so on. These and many other tasks often give rise to situations that demand a more generalized methodology than the derivation of a single straightforward measurement, but fall short of requiring the complexity and variety of processing indicated by the scene interpretation example. In such situations, an interpretation procedure generally adopts the techniques and methodology of the discipline known as pattern recognition.

In this chapter we shall examine some of the basic issues in pattern recognition and illustrate some approaches which might be considered, while subsequent chapters will take up the questions raised in a more formal and detailed way.

6.2 APPROACHES TO PATTERN RECOGNITION

In order to set the scene for a more detailed discussion of pattern recognition techniques, it is useful to consider some situations in which pattern recognition might be applicable in order to identify the particular difficulties which could be expected to occur in practice. To this end we will consider briefly three examples, all related in terms of the pattern types of interest, which collectively demonstrate the relationship between data characteristics and viable approaches, and the often encountered areas of discrepancy between theory and practice.

We shall take *character recognition* as the common focus of these preliminary examples, since this area offers well defined and easily described data. It is most important to be clear, however, that the formalized techniques introduced here are of much more general applicability and have been successfully used to interpret widely varying types of data. Character recognition is the area of application of pattern recognition involving the categorization of two-dimensional line drawing representations of characters – the most familiar example would be the reading of alphanumeric characters used in English text. A typical task for the designer of a character recognition system might be the specification of a device to read, say, postcode information on envelopes in the mail, or identification labels on cartons in a factory or warehouse.

With this in mind let us now consider as our first example a situation in which we are to design a system to read the numeric information typically printed along the bottom of a bank check, such as is shown in Figure 6.1. This information, usually giving details of sorting codes, account numbers and so on, would be useful in automating check handling processes or may be seen as one medium in a variety of code-reading tasks. (This sort of operation is often carried out using a reading system that detects magnetic ink, but for our present purposes it is quite reasonable to consider an optical system for the same sort of task.)

In fact, this sort of problem need not make many demands on the system designer – at least, not in the pattern recognition context – simply because the characteristics of the data are so clearly defined in advance.

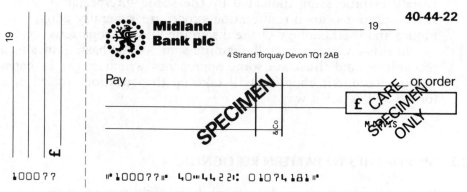

Figure 6.1 One application of character recognition is in automatic cheque sorting.

This sort of situation, where pattern archetypes are well defined and highly differentiated (the character type fonts are chosen specifically to meet these requirements, which is why they have that characteristic rather odd appearance) can generally be tackled by adopting a relatively unsophisticated 'mask matching' approach, where a character is compared with a set of masks each of which fits just one character shape precisely, until a match is found. Alternatively, again assuming the character shapes are chosen carefully, it might be possible to distinguish between a group of characters merely by computing the area of a character (the number of black pixels) within the chosen viewing area. Figure 6.2 shows a small group of (specially designed) characters which could be differentiated on this basis.

We have seen, therefore, that in the situation described an essentially *ad hoc* method of pattern recognition is adequate to tackle the problem, without resorting to any formalized technique or standardized procedure. It is important to begin at this level in order to see that the reason why such an informal approach is so effective here is attributable principally to the exceptionally favorable conditions which prevail. The character format

Figure 6.2 An example of a small set of stylized characters designed to constrain data variability.

is well defined and specifically selected in advance to maximize the differences between character types. Furthermore, the sort of environment in which this type of data is handled is generally also favorable, so that overall the data domain is highly constrained in a way which greatly simplifies the recognition task. Such conditions do not always prevail, as we shall see.

For our second example we shall consider a much less constrained area of operation, in which a system is required to read and recognize machine-printed English alphabetic characters generated from unspecified sources. In this case we have no *a priori* information about the precise characteristics of the data with which we have to work, since we have no control over the data generation mechanism (typewriters have many different typefaces) though, since we have specified 'machine-printed' data, we can assume a less widely varying set of examples to work with than would be expected if we allowed handwritten versions additionally.

In this situation the previous mask-matching approach is much less likely to be particularly effective, since we are not at all sure what our masks should look like. Consequently, we will try a slightly different approach which is much closer to a generalized model of a recognition procedure expressed in terms of its underlying mechanisms.

In this scheme we work from the starting point that it may be better to try and break down the overall patterns to be identified in some way, trying to find embedded features which, taken as a group, characterize a particular type of pattern. For example, with the particular sort of data chosen here we might make use of some straightforward geometrical features such as, for example, straight lines in either a horizontal or a vertical orientation, a right-angled junction between two straight lines, and so on. Such simple features are likely to be rather more easy to recognize than the much more complex global pattern, and an algorithm based on combining evidence about such individual features might be easier to envisage than some direct equivalent of the previous mask matching approach.

Table 6.1 *An analysis of some alphabetic characters using geometric feature descriptors*

Character	Features			
	Vertical straight lines	Horizontal straight lines	Oblique straight lines	Curved lines
L	✓	✓	—	—
P	✓	—	—	✓
O	—	—	—	✓
E	✓	✓✓✓	—	—
Q	—	—	✓	✓

If we can reduce a global problem to a series of smaller and more accessible sub-problems in this way, we might have a more viable means of automatically identifying the characters of interest. For example, if we define variables V, H, O and C and, for any character, assign to each variable a value indicative of the number of occurrences of the features specified as, say, vertical straight lines, horizontal straight lines, oblique straight lines and curved lines respectively, then we have the basis of a possibly more flexible recognition model.

To illustrate this approach, Table 6.1 shows a small subset of alphabetic characters described in terms of this small set of features which could be implementable within the framework of the proposed model, and it can then be seen that an algorithm such as that illustrated in Figure 6.3 might be an appropriate means of approaching the problem of identifying a character in relation to the possible categories offered in this example. Although this type of scheme represents a step forward from the mask-matching approach – in particular it is, in principle at least, capable of coping with a variety of styles and varying examples from within the pattern categories available – it soon becomes apparent that, in its present form, the approach has a number of potentially significant drawbacks. The following obvious difficulties should be noted explicitly and though this list is not exhaustive, the questions raised are of primary importance:

1. Is it possible to choose a set of features that gives a unique description of all the pattern categories available? Consider adding C to the list of Table 6.1, for example.
2. How easy is it in practice to detect the features that have been chosen? For instance, how 'horizontal' does a horizontal straight line need to be?
3. Is the definition of a particular pattern type always unambiguous? Look carefully, for example, at the feature-based definition of the P character. And so on.

Now it is clear that this particular model is not meant to be regarded as it stands as a directly implementable practical technique. Nevertheless, there are two aspects of its formulation which are valuable pointers to a more formally defined and practically applicable specification. The first is that it establishes the principle of defining a pattern in terms of its constituent 'features' which are the elements to be worked with in making a decision about the identity of the pattern, and this is an idea that will play an important part in later discussions. Secondly, the sequence of processes it implies – data sensing/coding, extraction of features, coordination of features, decision-making – is almost a canonical form of a formalized pattern recognition system formulation. The present discussion has consequently brought us much closer to the main objectives of pattern recognition methodology.

Before moving on it is interesting to consider a third pattern recognition situation. This example does not aim to introduce a further new

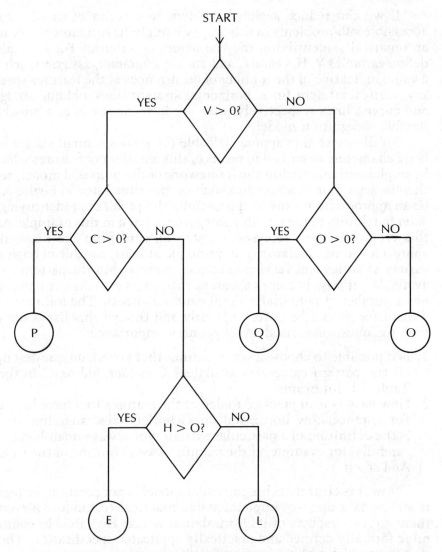

**Figure 6.3 A tree-structured algorithm for simple alphabetic character
identification.**

approach to the recognition problem, but rather illustrates how, even
though the approaches introduced do emphasize important principles, a
further degree of complexity generally arises when theoretical ideas are
translated into practical reality.

Hence, for our final introductory example, we will briefly consider a
practical application typified by the automatic reading of machine-printed
postcodes on envelopes in the mail. In Britain, this generally involves a
system design that will allow discrimination between 34 alphanumeric

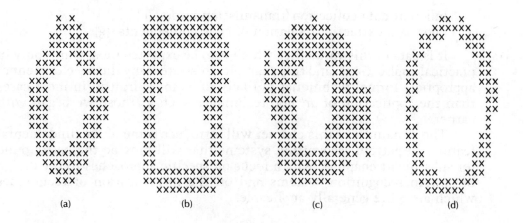

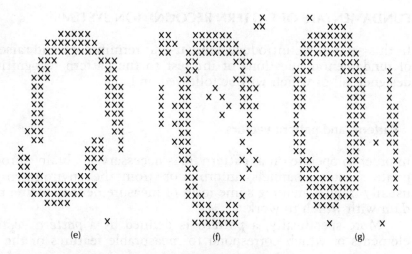

Figure 6.4 Examples of digitized machine-printed characters, illustrating typical intra-class data variability in a practical situation.

classes (the letters A–Z and the numerals 0–9, with the O/0 and I/1 pairs respectively assumed to be indistinguishable). The point at issue here is illustrated by reference to Figure 6.4, which shows some actual examples of typical data which occurs in such an application. These examples show binarized images of characters belonging to the O category. In particular it is striking that, bearing in mind that all seven examples belong to the same class, there is an enormous variability in the data. This can be traced to a variety of causes, including:

> Variability of type font, (a) and (b),
> Variability in character line thickness, (a)–(d),
> Variability in print quality/accumulated noise, (f),

Inherent data collection/transmission errors, (e),
Poor data extraction/isolation of code components, (g).

It is data of this variability which is to be expected most commonly in practical applications, and this example illustrates why the development of appropriate formal techniques is likely to be more fruitful in most cases than the application of *ad hoc* techniques such as those we began with earlier.

The remainder of this chapter will introduce some of the fundamental features of pattern recognition systems that will provide the background for subsequent chapters, which focus on specific approaches to the design of pattern recognition systems and on the formalization of techniques which are quite generally applicable.

6.3   FUNDAMENTALS OF PATTERN RECOGNITION SYSTEMS

In this section we introduce some useful terminology and raise a number of fundamental questions of interest to the pattern recognition system designer, all of which will be followed up later.

6.3.1   Patterns and pattern vectors

In order to operate on a 'pattern' it is necessary to obtain, through appropriate sensory channels, information from the environment, and this usually implies making some form of measurement to define the *pattern data* with which to work.

More specifically, a *pattern* is defined by a *pattern vector* ($\mathbf{X}$), the elements of which correspond to measurable features of the pattern or sensory data.

Thus,

$$\mathbf{X} = \{x_1, x_2, x_3, \ldots, x_n\}$$

is an n-dimensional vector where $x_1, x_2, \ldots, x_n$ are measurements of specified characteristics of the pattern, usually referred to as *features* or *pattern descriptors*.

At one extreme of simplicity these features may be equivalent to actual pixel values in a binarized image (see Figure 6.5), while in an alternative domain they could represent, say, the formant frequencies of a spoken sound (frequencies of high-intensity excitation caused by resonances in the vocal cavities). In the former case the pattern vector would take the form:

$$\mathbf{X} = \{00000101001000110000110001001010000 1\}$$

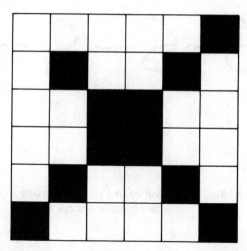

Figure 6.5 A low-resolution binarized X character.

while, in the latter, the form of the pattern vector would perhaps be:

$$\mathbf{X} = \{F_1, F_2, F_3, F_4\}$$

where F_n represents the nth formant frequency.

6.3.2 Pattern classes

The question of membership of a particular class or category of pattern is, of course, a preoccupation with the designer of a pattern recognition system, since the determination by computational means of just this sort of information is the goal he has in mind in the formulation of his design. A pattern class may be thought of as a group of patterns that share some common characteristics or attributes. It is important to realize that, in the real world, it is not always easy to determine class membership absolutely. This can be illustrated by reference to Figure 6.6, which shows almost a continuum of configurations with respect to a group of alphabetic characters which cause two initially distinct classes to merge as progressive distortions are applied to the archetype representations at each end of the spectrum.

It is a subtle question as to whether the example indicated belongs to the 5 or the S class. Furthermore, in this case there is further ambiguity since even two human observers might choose different classifications for this example. Presumably if a given individual wrote this character intending it to be a 5 that would fix its identity, though a consensus view among other observers might be that it is an S. This ambiguity in class membership can make life very difficult in some practical situations, particularly those where variability of form is an inherent data characteristic.

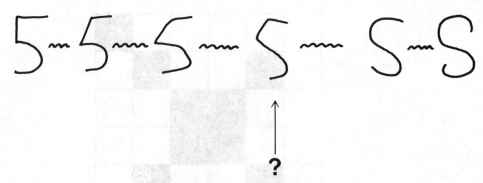

Figure 6.6 *Representation of two pattern classes as part of a continuous spectrum of sample forms.*

6.3.3 Appropriateness of features

We can pick up a further point which relates to both pattern vectors and pattern classes by asking some questions about features or attributes and their relation to class definition. The idea of choosing features to work with in a specific task is that they are chosen to contribute to the greatest possible extent to the process of identification of individual examples and to the discrimination between classes.

A short example will illustrate this point. Let us suppose that we wish to devise an automatic means of distinguishing between, say, bananas and grapefruit. Two physical measurements that might easily be made are weight and length. (We would have have to pin down the length concept a little here, perhaps by defining it as the maximum distance between two points in a cross-section of the fruit.) If we select these two parameters as the descriptors then, using a straightforward graphical representation (see Figure 6.7(a)), we can see that we have chosen wisely – it is likely that the two classes of interest will fall into reasonably distinguishable areas of the *feature space* defined by the length-axis and the weight-axis. Other possible measurements (for example, of colour) might not have proved quite so effective.

For this application, then, the descriptors chosen are adequate, but consider the problem that would arise if we were to try to use the same descriptors to separate, for example, grapefruit and oranges. Here the separability would be likely to be much less clear-cut, and we might have to consider substituting or adding other descriptors. (Colour would be much more helpful in this case, for example.)

By way of comparison, Figure 6.7(b) shows that, in the completely different data medium of speech, the use of the formant frequencies mentioned earlier (the first two in this illustration) can be relatively effective in

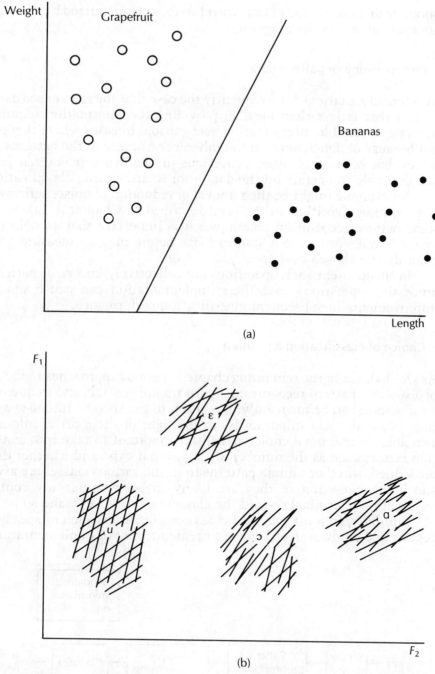

Figure 6.7 *Analytical characterization of a classification problem in two-dimensional feature space. ε = 'e' as pronounced in desk, u = 'u' as in rule, ɔ = 'o' as in fort, ɑ = 'a' as in card.*

phonetic discrimination of four vowel sounds characterized by the standard notation shown in the diagram.

6.3.4   Preprocessing of pattern data

In practical situations it is frequently the case that the raw sensed data is in a form that is less than ideal in providing the information required for accurate or reliable interpretation/recognition. In cases where this occurs, not because of deficiencies in the inherent structure of the patterns themselves, but because of other extraneous mechanisms, it is often possible and desirable to operate on the data prior to attempting classification.

An example might be the removal or reduction of noise, perhaps using a smoothing algorithm in the way described in Chapter 3. Likewise, for some pattern recognition techniques it is important that an object in an image is transformed to a standard size before making measurements to quantify the chosen descriptors, and so on.

In this context such operations are collectively known as pattern preprocessing operations, and their implementation can cause significant improvements in subsequent classification performance.

6.3.5   Choice of classification algorithm

As we shall see in the remaining chapters, various approaches to the design of a suitable pattern recognition algorithm are possible and frequently the 'best' choice can be made only in relation to the specific intended application area and, most importantly, in the light of the *a priori* information available. In making a choice it will be important to take into consideration factors such as the number of classes that exist and whether they are predefined, whether sample patterns from the various classes are available and how representative they are likely to be, whether any contextual information is available to aid the classification process, and so on.

Similarly, in most practical cases considerations such as speed of processing and hardware/software requirements, cost of implementation, and

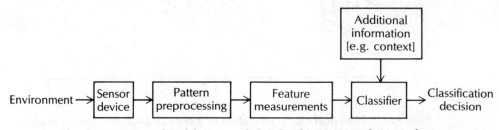

Figure 6.8 Generalized framework for the description of a complete pattern recognition system.

required level of performance will have to be taken into account. Thus the approach ultimately adopted will be highly problem-dependent. It is possible nevertheless to propose a general structure for the implementation of pattern recognition techniques, and this is shown in Figure 6.8. This schematic representation of a pattern recognition system will be used as a reference in later chapters.

6.4 SUMMARY

This chapter has introduced the basic concept of pattern recognition as an approach to the interpretation of commonly encountered image data. The examples have both introduced the underlying concepts and problems involved and pointed to a need for a formalization of problem areas and methods of solution. The requirements of a typical pattern recognition system have been introduced and prepare the way for the basic mathematical treatment of the following chapters.

It can already be seen how classification/interpretation of visual data can rely extensively on the ideas and techniques of image processing described earlier. It is important, for instance, to be able to make quantitative measurements on imaged objects in order to generate pattern descriptors to work with in a recognition task, preprocessing to maximize classification ability requires basic image processing in order, perhaps, to normalize image data or to emphasize characteristics that might contribute to the recognition procedure, and so on. Although we shall now move on more to deal with the methodology of pattern recognition, this very important link with earlier chapters should not be overlooked.

6.5 EXAMPLES FOR SELF-ASSESSMENT

6.1 Discuss the problems involved in extending the set of characters shown in Table 6.1 (see text) to include the whole set of symbols A–Z, and suggest how the range of features used might be extended.

Illustrate your discussion with examples where appropriate.

6.2 Table 6.2 shows two pattern classes, A and B, where individual patterns are represented in terms of patterns descriptors x_1 and x_2. Find, graphically or otherwise, the equation of a discriminant function to separate these two classes.

Discuss the behavior of a pattern dichotomizer (a two-way discriminator) implemented according to the specification you have described, if an error in the data collector is such that the value of each measured x_2 descriptor is artificially reduced to four-fifths of its correct value.

Table 6.2

Pattern number	Class A		Class B	
	x_1	x_2	x_1	x_2
1	30	25	30	15
2	35	75	40	20
3	40	60	45	30
4	35	45	55	30
5	35	35	65	35
6	45	40	55	40
7	45	55	65	45
8	50	65	70	25
9	55	70	80	20
10	60	60	70	55
11	55	50	85	45
12	60	65	82	22

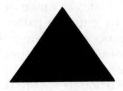

Figure 6.9

6.3 Figure 6.9 shows binarized images of four components in an industrial assembly task (drawn approximately to scale) which are to be identified by machine. Explain in some detail how appropriate information may be extracted from images of these objects to allow discrimination between them.

6.4 By means of examples, show how the image processing techniques introduced earlier might be used in appropriate preprocessing to improve the possibility of automatic classification of images.

7 ★ Formal methods in pattern recognition

7.1 INTRODUCTION TO STATISTICAL PATTERN ANALYSIS

In order to provide a background to the generalized theory that follows, we shall begin the more detailed discussion of pattern recognition theory by considering a somewhat constrained example, but one, nevertheless, that embodies the main problems the theory must address.

Let us imagine a plant manufacturing nuts and bolts such as those shown – approximately to scale – in Figure 7.1. We will assume a situation in which these components are carried along a conveyor belt, in a randomly occurring order, and specify that it is required to identify each component as it appears at a given point so that some further action (such as packing or, perhaps, utilization in an assembly task) can be initiated. We will assume that components appear at the point of interest randomly one at a time, and that at some earlier stage along the line a passive process (perhaps feeding through a narrow aperture) has ensured that the components are all lying flat on the belt to give a top-view profile corresponding to that already shown in Figure 7.1.

The problem can be summarized as follows, referring to Figure 7.2. At some particular time a single component will appear at the sorting point, and on the basis of any information available at the time we wish to determine whether the component is a nut (which we will designate a class C_1 component) or a bolt (referred to as a class C_2 component).

The statement '... on the basis of any information available at the time ...' is very important, since the extent and type of 'knowledge' about the problem will play a significant part in determining both general strategy and actual level of performance, and this is a concept to which we shall regularly return.

In the simplest sort of decision-making procedure the amount of information available could be extremely rudimentary. Perhaps we cannot actually inspect the components at all as they appear at the sort point, and all we know is that they are manufactured according to some process that allows us to predict the relative numbers of each class that are likely to appear on the conveyor belt at any time. (Although not important here, it is as well to bear in mind that even this information would be subject to variation over a period of time, and we would need to acquire some knowledge of this as well.)

In this sort of situation (admittedly not very practical), the best strategy

(a)   (b)

Figure 7.1 Two 'archetypal' class-defining samples: (a) bolt, class C_2,
(b) nut, class C_1.

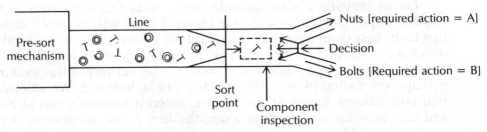

Figure 7.2 Representation of a possible problem organization in an
industrial inspection task.

we could adopt to identify an unknown component would be to 'guess' on
the basis of the relative probability of occurrence of the two component
type. Hence we would make a decision by using knowledge only of $p(C_1)$
and $p(C_2)$, the *a priori* occurrence probabilities with respect to class C_1
(nuts) and class C_2 (bolts) respectively. Of course, in practice, this is a
very contrived and unlikely situation but, even where we are rather more
sophisticated, any decision rule we adopt will in principle still be based on
a guess, though we would hope to evolve a basis for choosing decision rules
that give increasing confidence in this seemingly tentative approach.

In a realistic situation, then, we feel that we need more knowledge –
more information to help us to make a better judgment in formulating a
decision rule. Now, of course, it is almost always the case that we can
obtain further information, because it is most likely that we can make
some measurement on an arbitrary component to help us to decide on its
identity or class membership. Different measurements would require different measuring techniques, but some likely candidates which might
be considered could be things like object area, maximum object length,
weight of object, the existence of a hole within the object surface, and
so on. Any or all of these or, indeed, other measurements might be appro-

priate in helping to identify an object, and the set of measurements made would correspond to the pattern descriptors discussed in Chapter 6.

Let us return to the specific example under consideration and assume that we have positioned some suitable visual imaging device at the sort point, arranged so that it produces binarized images such as those of Figure 7.1. Let us further assume that we have chosen to measure object area to try to help in the decision-making process, as this will be an easy parameter to derive from the raw image data. We now have available exactly the sort of additional information discussed earlier, which we ought to be able to exploit. However, this can only be utilized effectively if we have some additional knowledge about the current situation. In this case, the knowledge required would be knowledge about the relationship between object area and membership of class C_1 and class C_2 (i.e. the extent to which this measurement is sufficient to describe nuts and bolts).

Let us suppose that we have such knowledge available, embodied in the form of a probability distribution function with respect to object area for each of the two classes of interest. These distributions would be designated 'class conditional' distributions and might assume the form shown in Figure 7.3. These distributions contain an inherent description of the object classes of interest (with respect to the parameter we have chosen to measure) and, moreover, we should be able to acquire some information

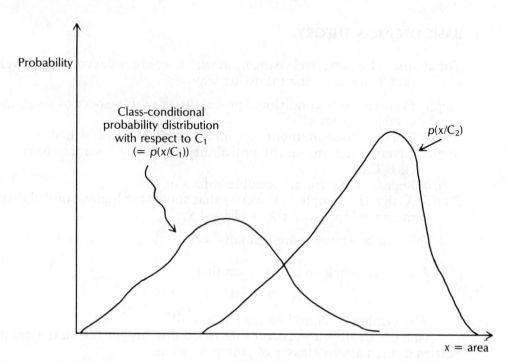

Probability

Class-conditional
probability distribution
with respect to C_1
$(= p(x/C_1))$

$p(x/C_2)$

x = area

Figure 7.3

about their characteristics from a study of (known-identity) samples. They illustrate again, however, how in practical situations pattern recognition can be a difficult problem; even though we have specifically identified a parameter we feel will be helpful in discriminating between the available classes, it is by no means a straightforward matter to carry out the classification perfectly. As Figure 7.3 shows, individual instances of class membership can show a significant variation in the measured parameter arising from a variety of sources such as allowed tolerances in the manufacturing process, tool wear, perhaps noise in the imaging system or measuring technique, and so on. Likewise, the overlap between the two class-conditional distributions shows that in some cases it will not be possible unambiguously to identify a component purely on the basis of the parameter chosen.

Despite this, if the knowledge contained in these distributions is a good representation of the statistical characteristics of the data classes, it should be possible to exploit this information in such a way that we make the best guess possible when trying to identify an unknown object on the basis of an observed measurement. This is the underlying motivation of statistical pattern classification algorithms, which we shall pursue in more detail in the following section.

7.2 BASIC DECISION THEORY

The approach for pattern classification which we are endeavoring to develop can be sketched out in the following way:

Step 1 Evaluate class conditional probabilities with respect to some measurable feature x.

Step 2 Obtain a measurement of x for a sample to be classified.

Step 3 Given x, determine the probability that such a sample belongs to class C_i.

Step 4 Repeat step 3 for all possible values of i.

Step 5 Assign the sample to the class that shows the highest probability of membership, given the measured x.

This can be stated more formally as:

Assign the sample to class C_i such that

$$p(C_i/x) > p(C_j/x) \qquad \forall \, j \neq i$$

(The symbol $\forall$ should be read as 'for all'.)

This constitutes a *decision rule* based directly on the statistical properties of the pattern classes of potential interest.

Unfortunately it is not generally very easy to discover precisely the required distribution $p(C_i/x)$. This is principally because if we are to determine accurately a value for $p(C_i/x)$ it would be necessary to obtain and observe a very large number of pre-identified samples for each x-value that can occur, a situation which is usually quite difficult to contrive in most practical cases. In the example we have been considering, for instance, we have to be content with known-identity samples which are available from the output of the manufacturing process and, though we are at liberty to be selective in what we retain, we often have relatively little control over the initial training set available.

It is more practical, at least in principle, to use Bayes theorem to make this decision rule specification more viable, and consequently we may wish to use instead the alternative form:

$$p(C_i/x) = \frac{p(x/C_i)p(C_i)}{p(x)}$$

where

$$p(x) = \sum_{\forall i} p(x/C_i)p(C_i)$$

We note that $p(x)$ here is a scale factor and does not influence the relative conditional probability values, and can therefore be ignored. Hence, the alternative form of the statistical decision procedure may be expressed in terms of the rule:

Assign the sample to class C_i such that

$$p(x/C_i)p(C_i) > p(x/C_j)p(C_j) \qquad \forall j \neq i$$

Two important points emerge from this discussion:

1. $p(x/C_i)$ is, in principle, much easier to estimate from a set of known-identity samples than would be the case with $p(C_i/x)$, although some remaining difficulties will still be apparent in practice.
2. In this form, the decision rule offers a procedure that will minimize the probability of misclassification, provided that the conditional probability density functions are known accurately.

This latter point is extremely important in pattern recognition theory. Reference back to Figure 7.3 for the two-class case will show this intuitively, as implementing the rule will place the decision boundary between C_1 and C_2 at the intersection point between the two distributions, thus minimizing the area within which erroneous classification can occur. This decision rule is known as the *Bayes decision rule*, and a *Bayes classifier* based on an implementation of the rule is a useful yardstick against which to compare the performance of other classification schemes.

Unfortunately, there is a degree of oversimplification implicit in the preceding paragraph. In practice, the optimal Bayes performance can be achieved only if the associated probability distribution functions are accurately known, while the nature of practical pattern recognition is, of course, that this is rarely the case. Consequently, most practical classifiers represent only an approximation to this ideal. Indeed, it is fair to say that, as a general rule, the performance of a statistical classifier is only as good as the estimation obtained for these probability distributions, with the result that huge efforts have been made over the years to develop methods of obtaining good estimates while maintaining computational effort within practically manageable bounds. This is a problem to which we shall be returning shortly.

7.3 GENERALIZED DECISION THEORY

We shall now attempt to broaden the preliminary concepts outlined in the previous sections to provide a more general view of pattern recognition as a statistical process involving formal decision theory.

In particular we shall consider making measurements to define a set of pattern descriptors $\mathbf{X} = \{x_1, x_2, \ldots, x_N\}$ and work within an environment in which we can identify n classes $C_1, C_2, \ldots, C_n$.

Furthermore, it is useful to consider the mechanisms of decision making in the context specifically of pattern classification. Let us restate the fundamental problem of classification in its simplest form as it relates to our particular concerns. (It is convenient to refer to Figure 7.4 at this point.)

A sample pattern from some (*a priori*) unknown class C_i, characterized by a feature vector $\mathbf{X}$, is presented to the classifier. The classifier will implement some algorithm which, on the basis of the component values in $\mathbf{X}$ and its inbuilt knowledge (supplied by the system designer or acquired in some other way), will compute and generate a 'decision', assigning the pattern $\mathbf{X}$ to some class C_j. If, however, we *do* have some initial information about the class to which $\mathbf{X}$ actually belongs, we can draw some conclusions about the way in which any adopted decision rules are working.

It is clear that at least two possible situations might prevail, as follows:

1. If $C_j = C_i$, then the classifier has made a *correct* decision with respect to $\mathbf{X}$.
2. If $C_j \neq C_i$, on the other hand, the classifier has generated an erroneous decision. We say that a *substitution error* has occurred, and the un-labeled sample is given a label corresponding to an inappropriate class.

It may be noted at this point that there is a further possibility, namely that no definite class assignment can be made. This situation may be considered as providing the possibility of a *rejection* decision from the

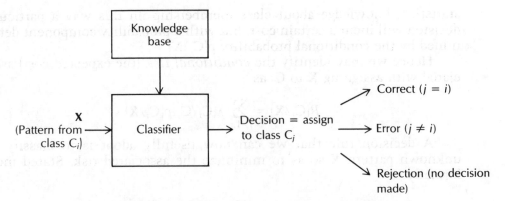

Figure 7.4 General classifier decision structure.

classifier. This is a case we shall meet again later, but for the time being the real point of significance is the fairly obvious one, that in making a decision at all there is the possibility of error. More generally, the making of a decision carries with it an associated *cost* (sometimes called a *loss*) which may be characterized as

$$\lambda(C_i/C_j)$$

and which is to be regarded as a measure of the cost incurred for assigning **X** to class C_i when, in fact, it belongs in reality to class C_j.

The idea of associating a cost function with classification decisions allows great flexibility in determining decision functions and specifying classification algorithms, which can then be adapted to reflect the important features of a particular situation. For example, in an automatic inspection system for a manufacturing process (where perhaps we wish to make a visually based quality inspection of the nuts and bolts in the process with which we have been illustrating the ideas of this chapter, and either accept (pass) or discard (fail) them), we may feel that the cost of accepting a faulty part should be rather greater than if we fail to accept a perfectly good component. This may prove to be the cheapest option in the long run when the whole manufacturing/marketing operation is considered. The idea of a cost function allows us to incorporate just this sort of information into the classification process.

Conversely, we may decide that all errors should be equally costly, in which case we would be likely to choose a particularly simple form of cost function such as:

$$\lambda(C_i/C_j) = \begin{cases} 0 & \text{for } i = j \\ 1 & \text{for } i \neq j \end{cases}$$

The question remains, however, of utilizing this information. The most obvious approach is to couple the cost information with the basic statistical knowledge about class membership. In this way a particular decision will incur a certain cost, but with a probability component determined by the conditional probability $p(C_i/\mathbf{X})$.

Hence we may identify the *conditional risk* (the expected cost) associated with assigning $\mathbf{X}$ to C_i as

$$R(C_i/\mathbf{X}) = \sum_{j=1,n} \lambda(C_i/C_j)p(C_j/\mathbf{X})$$

A decision rule that we can now usefully adopt is to classify an unknown pattern $\mathbf{X}$ so as to minimize the associated risk. Stated more formally:

Assign $\mathbf{X}$ to C_i such that

$$R(C_i/\mathbf{X}) < R(C_j/\mathbf{X}) \qquad \forall\, j \neq i$$

Furthermore, if we were to apply this rule to all possible decisions – i.e. for every possible $\mathbf{X}$ – we would minimize the overall risk involved in making a classification. Consequently, this decision rule represents the best performance that can be achieved, and, from a slightly different viewpoint, the Bayes classifier is seen to represent a theoretically optimal formulation. It is readily seen that the idea of minimizing risk ties in directly with the idea of optimal error performance, and this may easily be shown (and is left as an exercise for the reader) directly by considering the binary cost function introduced earlier.

One further general point should be mentioned here, which is particularly important in relation to the sort of decision rules we have been suggesting. In general, we are proposing a rule of the form:

For a given $\mathbf{X}$ compute an appropriately formulated decision function $d_i(\mathbf{X})$ for each class $C_i (i \in \{1, 2, 3, \ldots, n\})$.
Then, $\mathbf{X} \in C_i$ such that

$$d_i(\mathbf{X}) > d_j(\mathbf{X}) \qquad \forall\, j \neq i$$

We have sketched out a possible general form for the functions $d_i(\mathbf{X})$, though without discussing their derivation in much detail at this stage. It is worth noting, however, that this generalized decision rule defines a standardized or 'canonical' form of a classification scheme in which the precise nature of the $d_i(\mathbf{X})$ functions may be determined according to a particular situation. More to the point for the present, however, is the fact that the very nature of this scheme is such that it contains some potential hazards in practical situations.

The scheme is summarized diagrammatically in Figure 7.5. In response to $\mathbf{X}$ the decision function $d_i(\mathbf{X})$ is computed to give a quantitative measure of the association of $\mathbf{X}$ with the class C_i, and repeated for all classes. The overall classification decision is based on selecting the maximum value from the i values so generated.

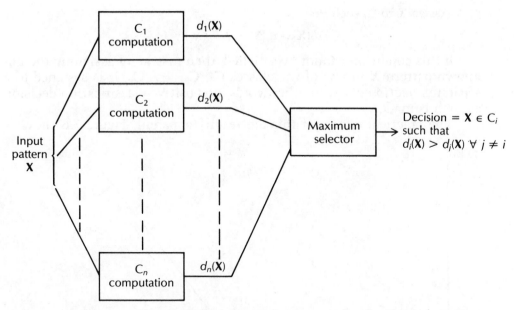

Figure 7.5 Decision functions in the context of a classification system.

Now suppose for a moment that the nature of the $d_i(\mathbf{X})$ computation is such that its value for any i can be represented on a magnitude scale from 0 to 100. We may then view the value observed from any $d_i(\mathbf{X})$ computation as representing a sort of percentage association between the pattern sample $\mathbf{X}$ and the ith class, as measured by the decision-making computation. Thus, the higher this figure is for a particular class, the more likely it is that $\mathbf{X}$ will be classified as one of its members. But here a practical drawback becomes apparent. Let us suppose that in response to a particular pattern $\mathbf{X}$ we find that $d_l(\mathbf{X})$ generates the highest such score while $d_k(\mathbf{X})$ generates the next highest. Provided that $d_l(\mathbf{X}) - d_k(\mathbf{X})$ is relatively large, we can be reasonably confident about our decision – at least in relation to the specific functions chosen – but if $d_l(\mathbf{X}) - d_k(\mathbf{X})$ is small (say less than about 5%) we can be much less confident that our decision is right, since we are now making the classification on the basis of marginal evidence.

For this reason we might often find it helpful to introduce into a classification algorithm the idea of a *rejection margin* or *confidence level* (γ). There are two similar ways in which we can use such an idea. The first is that, in applying the original rule directly, γ can be evaluated as the (possibly scaled) difference $d_l(\mathbf{X}) - d_k(\mathbf{X})$ as defined above, thereby allowing us to attach a relative quantified degree of confidence to each classification decision made.

The second, and more common, method of utilizing confidence information is to incorporate a rejection margin γ directly into the decision rule, which then becomes:

Assign **X** to C$_i$ such that

$$d_i(\mathbf{X}) > d_j(\mathbf{X}) + \gamma \qquad \forall\, j \neq i$$

If this condition cannot be satisfied, then instead of assigning the unknown pattern **X** to one of the classes C$_1$, C$_2$, ..., C$_n$, it is assigned to a separate *rejection class*. In other words, no positive classification decision as such is made.

The practical effects of this are readily apparent. Figure 7.6 shows a

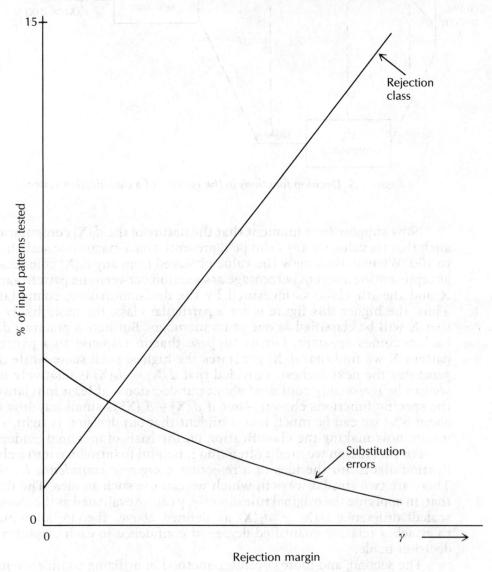

Figure 7.6 Effects of a rejection margin in a typical classification task.

typical performance (actually derived from a recognition experiment with postcode characters on envelopes in the mail). Application of the original decision rule corresponds to the case when $\gamma = 0$. The effect of increasing γ (effectively the same as requiring a 'stronger' decision to be made) is primarily to reduce the number of errors made (since most errors are likely to arise as a result of marginal decisions) at the expense of declining to make any positive decision at all about a generally more rapidly increasing number of samples which are consequently rejected according to the definition above.

The same idea is applicable in an extension of this scheme to another situation in which errors would be particularly likely to occur – where even the *maximum* value of a decision function obtained is relatively low on the percentage scale. This would imply that the unknown pattern $\mathbf{X}$ is unlikely to belong to any of the predicted available classes (perhaps, in the running example chosen, it is some stray object among the nuts and bolts on the conveyor belt). Thus a lower bound ($d_{\text{threshold}}$) on the decision function 'score' could be imposed in the classification rule such that rejection occurs if the maximum decision function value $d_{\max}(\mathbf{X})$ falls below this threshold (i.e. if $d_{\max}(\mathbf{X}) < d_{\text{threshold}}$), with an effect on error rate performance similar to that in the case previously discussed.

This rejection mechanism adds yet another dimension of flexibility to the design of a pattern recognition system, allowing the system designer to control a trade-off between error rate and rejection rate. A very important factor here, however, is that, while errors are difficult to recover in many situations, rejections can often be reprocessed in some alternative way (e.g. by human inspection if necessary), thereby increasing the overall correct recognition rate of the system.

7.4 ESTIMATION OF CLASS-CONDITIONAL PROBABILITY FUNCTIONS

As we have already discovered, the performance of a statistical classifier will depend crucially on the estimation of the class conditional probabilities $p(\mathbf{X}/C_i)$.

In principle, this estimation should be straightforward, and should rely on measurement of the elements x_i of the vector $\mathbf{X}$ for a large enough set of known-identity samples (a training set) to give statistically accurate results. Of course, this point in itself identifies one area of some difficulty – we generally have less than total control over the composition of the training set and it may be difficult to accumulate a statistically representative set of samples. Equally, we theoretically need to obtain samples with a distribution of feature measurements $\mathbf{X}$ for all possible $\mathbf{X}$-vectors that we are likely subsequently to encounter. This is an improbable task in a realistic situation, and in any case we can never be entirely confident in this general area, since it is precisely because we cannot know exactly what patterns are going to appear that we are designing the system in the first place.

One way of approaching this problem is to begin by making an assumption about the overall *form* of the distribution. If the form chosen represents a reasonable assumption, we should get a reasonable level of performance from a classifier based on this assumption and, furthermore, our principal computational problem will then reduce to that of determining the parameters that define the specific distribution with the general form chosen. We shall illustrate this idea both to demonstrate a practical approach to statistical classifier design and because it will allow us to make a number of further theoretical generalizations about the underlying methodology of pattern recognition.

Let us assume that the $p(\mathbf{X}/C_i)$ distribution is reasonably represented by assuming a normal distribution. (This is, in fact, quite a reasonable assumption in many practical situations.) Then, since we are dealing with a multidimensional feature space within which a pattern $\mathbf{X}$ is found, we may describe $p(\mathbf{X}/C_i)$ by the multivariate normal distribution function.

Thus,

$$p(\mathbf{X}/C_i) = (2\pi)^{-n/2}|\Sigma_i|^{-1/2}\exp[-\tfrac{1}{2}(\mathbf{X} - \mu_i)'\Sigma_i^{-1}(\mathbf{X} - \mu_i)]$$

where μ_i is the vector of means for each element of $\mathbf{X}$, Σ_i is the ith class covariance matrix;

$$\Sigma_i = \begin{bmatrix} \sigma_{i11} & \sigma_{i12} & \cdots & \sigma_{i1N} \\ \sigma_{i21} & \sigma_{i22} & \cdots & \\ & & \ddots & \\ \vdots & \vdots & \sigma_{i1N} & \vdots \\ & & & \ddots \\ \sigma_{iN1} & \cdots & & \sigma_{iNN} \end{bmatrix}$$

in which σ_{ikl} is the average of $(x_{mk} - \mu_{ik})(x_{ml} - \mu_{il})$ over the available training samples (m) and represents the degree of correlation between the kth and the lth elements of $\mathbf{X}$ in the ith class patterns (zero where x_k and x_l are statistically independent); and N is the dimensionality of $\mathbf{X}$.

The assumption of a multivariate normal distribution envisages a situation where pattern samples form a defined volume or cluster in multidimensional feature space, where the cluster center depends on the vector of means of the x_i measurements and where the shape of the sample cluster will be determined by the values of the elements of the covariance matrix.

If we take two-dimensional feature space, for example, we can easily visualize the situation by sketching in lines (similar to isobars on a weather map) which connect points of equal probability density, and it will be seen that these are elliptical in shape (see Figure 7.7). Visualizing the situation in this way and examining the mathematical form of the decision function

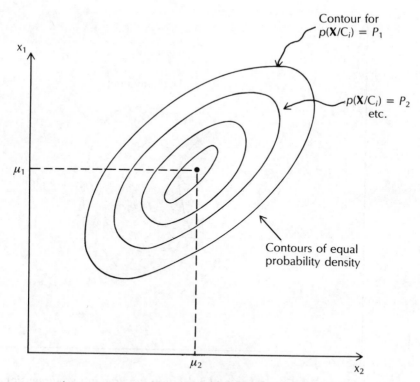

Contour for
$p(\mathbf{X}/C_i) = P_1$

$p(\mathbf{X}/C_i) = P_2$
etc.

Contours of equal
probability density

Figure 7.7 *Two-dimensional probability density contours.*

will enable us to understand the classification process in rather more detail. We shall first of all consider the most general case and then see how under certain further constraints other features of the classification process emerge.

For convenience we shall examine the discriminant function in a logarithmic form, such that

$$d_i(\mathbf{X}) = \log p(\mathbf{X}/C_i) + \log p(C_i)$$

Hence, under the conditions discussed above, we may write

$$d_i(\mathbf{X}) = -1/2(\mathbf{X} - \boldsymbol{\mu}_i)'\Sigma_i^{-1}(\mathbf{X} - \boldsymbol{\mu}_i) - 1/2\log|\Sigma_i| + \log p(C_i)$$

where we have dropped the term $-N/2\log 2\pi$ on the basis that it is not class-dependent and does not therefore contribute to the classification process.

If this expression is expanded to give

$$d_i(\mathbf{X}) = -1/2[\mathbf{X}'\Sigma_i^{-1}\mathbf{X} - 2\Sigma_i^{-1}\boldsymbol{\mu}_i\mathbf{X} + \boldsymbol{\mu}_i'\Sigma_i^{-1}\boldsymbol{\mu}_i] + \log|\Sigma_i| + \log p(C_i)$$

we can see that the form of the decision function is quadratic, an observation which is confirmed by reference again to the bivariate case (Figure 7.8). This representation, restricted to two classes for illustration, also

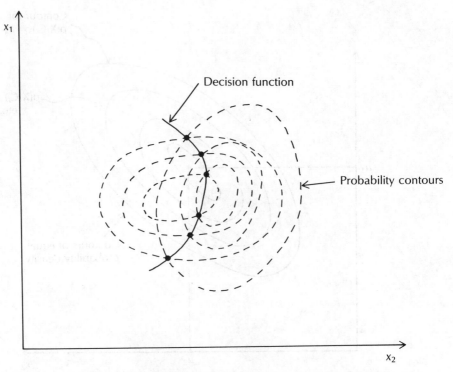

Figure 7.8 General case of a decision function in two-dimensional feature space.

demonstrates that the decision function splits the feature space into two regions, each characterizing one of the two possible classes to which **X** could belong, the boundary being determined by the line that passes through the points of intersection of the lines of equal probability with respect to the two class distributions.

If the parameters that specify the distribution – namely the mean vectors and the covariance matrix elements – are calculated on the basis of an available training set of sample patterns, then the decision rule already derived can be applied to give an implementation of a statistical classification algorithm.

However, under more favorable conditions, the decision rule can be simplified considerably and, additionally, can offer further insight into the nature of the classification process. We shall consider two special cases.

7.4.1   Special case (i)

In a particularly favorable problem domain where it is reasonable to assume identical covariance matrices for every pattern class such that:

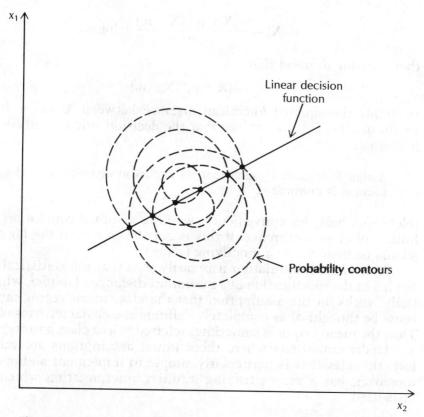

Figure 7.9 Decision function for special case (i) considered in the text.

1. the features chosen are statistically independent (i.e. the covariance matrix for C_i has elements $\sigma_{ikl} = 0 \; \forall \; k \neq l$), and
2. the variance of each feature is the same (σ^2),

then a great simplification in classification rule can be effected, as the form of the decision function (ignoring all the terms that are class-independent) reduces to

$$d_i(\mathbf{X}) = \frac{1}{2\sigma^2}(2\mu_i'\mathbf{X} - \mu_i'\mu_i) + \log p(C_i)$$

In this form it is seen that the decision function is *linear*, and again this can be confirmed by sketching out its relation to the class-conditional distribution functions (now transformed under the constraints noted from elliptical to circular shape) as shown in Figure 7.9.

Moreover, since the simplified expression above for $d_i(\mathbf{X})$ is a re-arrangement (with the $\mathbf{X}'\mathbf{X}$ term ignored as being class-independent) of the expression

$$d_i(\mathbf{X}) = \frac{-(\mathbf{X} - \mu_i)'(\mathbf{X} - \mu_i)}{2\sigma^2} + \log p(C_i)$$

then, bearing in mind that

$$(\mathbf{X} - \mu_i)'(\mathbf{X} - \mu_i)$$

is simply the squared Euclidean distance between $\mathbf{X}$ and μ_i (the mean vector of class C_i), we can see that the decision rule is equivalent to the following:

> Assign $\mathbf{X}$ to class C_i such that its Euclidean distance from the class mean μ_i is minimized.

(Note that here, for convenience, we have assumed equal *a priori* probabilities of class occurrence. If this is not the case then the $\log p(C_i)$ term should be included as a modifying factor.)

From this reformulation it is easily seen that the statistical approach has led to the specification of a *minimum-distance classifier*, which essentially works on the assumption that the class mean vector can in some sense be thought of as completely defining the characteristics of its class. Thus the mean vector is sometimes referred to as a class *prototype* pattern.

Under conditions where these initial assumptions are valid, therefore, this classifier is particularly simple to implement and operate and, moreover, has a very satisfying intuitive interpretation which is easily visualized.

7.4.2 Special case (ii)

For this commonly considered case we assume rather less stringent constraints, namely only that the class covariance matrices are equal. Thus the distribution contours are elliptical and of equivalent shape, but positioned according to the mean vector values. Figure 7.10 shows that in this case the decision function again has a linear form and this is apparent from the mathematical specification since, ignoring terms that are class-independent, the form of $d_i(\mathbf{X})$ reduces to:

$$d_i(\mathbf{X}) = -1/2(\mathbf{X} - \mu_i)'\Sigma^{-1}(\mathbf{X} - \mu_i) + \log p(C_i)$$

The expression

$$(\mathbf{X} - \mu_i)'\Sigma^{-1}(\mathbf{X} - \mu_i)$$

is also a measure of distance, and defines the square of the so-called Mahalanobis distance between $\mathbf{X}$ and μ_i.

Consequently in this case we again have a simple distance-based classification rule which may be stated as:

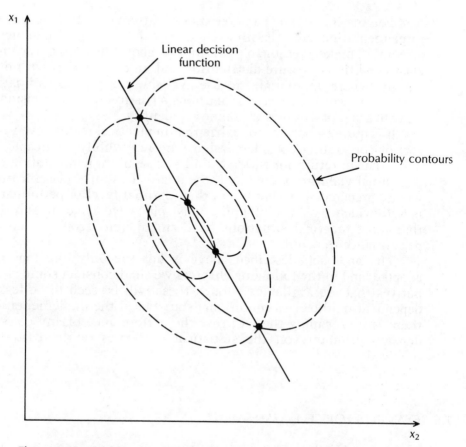

Figure 7.10 *Decision function for special case (ii) considered in the text.*

Assign **X** to C_i such that its Mahalanobis distance from the class mean μ_i is minimized.

As before, non-equal *a priori* class occurrence probabilities will modify this rule by the inclusion of the $\log p(C_i)$ factor.

Not only does this special case provide another rather simplified means of implementing a classification rule which is again readily grasped, but it is also a useful reminder that the concept of 'distance' is not restricted to any one measure but can be interpreted fairly broadly.

7.5 SUMMARY

In this chapter we have seen how decision theory can provide a basis for the design of a classification scheme which utilizes knowledge of a situation based on the statistical characteristics of pattern classes. One way in which implementation of a classifier for practical use might be achieved is to attempt to make a reasonable assumption about the form of the underlying class conditional feature distribution and, in conjunction with other constraints where appropriate, to use available samples to estimate a parameterized form for these distributions so as to define a decision rule for classification of samples of unknown identity.

It is quite important to realize that in practical reality, even where the sort of constraints considered above are not wholly applicable, we may judge that in return for the tractability, ease of implementation and computational viability offered by these relatively simple classification rules, we are prepared to accept a less than optimal level of performance. This is self-evidently a useful philosophy, since we have already indicated that most practical situations have an inherent scope for cost against performance trade-off.

The approach described above is not the only one that might be adopted, and in the following chapters we shall consider some of the alternatives that are available. Though these will be seen to be less directly dependent on the class conditional statistics at the implementation stage, there is no escaping the fact that the performance of any classifier will depend ultimately on the statistical properties of the pattern classes involved.

7.6   EXAMPLES FOR SELF-ASSESSMENT

7.1   Show that the adoption of a binary cost function in a statistical classifier produces an equivalence between minimum risk performance and minimum error-rate performance.

7.2   In a recognition task involving binarized patterns a number of known-identity samples from each class of interest are available as a training set in deriving an appropriate classification rule.

In a recognition task involving binarized patterns a number of known-identity samples from each class of interest are available as a training set in deriving an appropriate classification rule.

If it may be assumed that pixel values can be used directly as pattern features and that in this particular case these features are statistically independent, derive an appropriate expression for a decision function for use in a statistical classification procedure to identify the patterns.

Show that this decision function has a linear form.

7.3   In a certain factory process an automated visual inspection system is used to classify product samples as either Grade I or Grade II, and involves the measurement of two descriptors x_1 and x_2. A study of labeled samples of each grade shows the descriptor measurements in each case to be approximately normally distributed and characterized by the data shown in Table 7.1 where Σ_i represents the ith distribution covariance matrix and $|\Sigma_i|$ its determinant.

Table 7.1

	Grade I samples		Grade II samples	
Mean of x_1 values	120.6		94.2	
Mean of x_2 values	7.8		14.8	
Σ^{-1}	$\begin{bmatrix} 0.0021 & -0.012 \\ -0.012 & 0.144 \end{bmatrix}$		$\begin{bmatrix} 0.007 & 0.069 \\ 0.069 & 1.141 \end{bmatrix}$	
$\lvert\Sigma\rvert$	4711.2		552.3	

Stating any assumptions made, formulate a decision rule that could be used to determine the identity of further samples and hence determine to which grade a sample with descriptor values for x_1 and x_2 respectively of 105·8 and 9·1 should be assigned.

Briefly describe a means of reducing the probability of erroneous classification and comment on the practical implications of doing this.

8 ★ Alternative strategies for classifier design

8.1 INTRODUCTION

In Chapter 7 we examined an approach to the classification of patterns which was based directly on the statistical characteristics of the pattern classes involved and the extrapolation from the specific properties of a (restricted) training set of samples to the general properties of pattern classes which these represent. Thus, the decision functions employed for classification consisted of, or were derived directly from, statistical likelihood estimates for class membership.

An alternative, which will be considered first in this chapter, is to adopt an approach which focuses principally and directly on the form of the decision function, rather than as derivative from statistical considerations. In other words, rather than generating class-membership information by assuming that the underlying statistical feature distributions are of a known form, we can now consider an approach which instead makes an assumption about the form of the decision function to be used. We shall see later that this need not be regarded as a completely disjoint viewpoint, though it is clear that it involves a rather different conceptual framework.

Let us begin by considering a simplified case, where two pattern classes are to be represented in terms of just two features, such that $\mathbf{X} = \{x_1, x_2\}$. Let us suppose that for an available sample set of data we are able to plot individual pattern examples in feature space, as illustrated in Figure 8.1. By inspection, we can see that it is possible to define a decision surface that can easily be used to separate out or discriminate between the two pattern groups. It is also possible to work out a mathematical statement of the form of this decision boundary, as the equation of the dividing line between the groups as shown is seen to be

$$x_1 = x_2^2 - 5x_2 + 10$$

This mathematical description of the boundary between classes can then be used to formulate a decision rule. First, we can define a decision function $d(\mathbf{X})$, such that

$$d(\mathbf{X}) = x_1 - x_2^2 + 5x_2 - 10$$

and note (see Figure 8.2) that the condition

$$d(\mathbf{X}) = 0$$

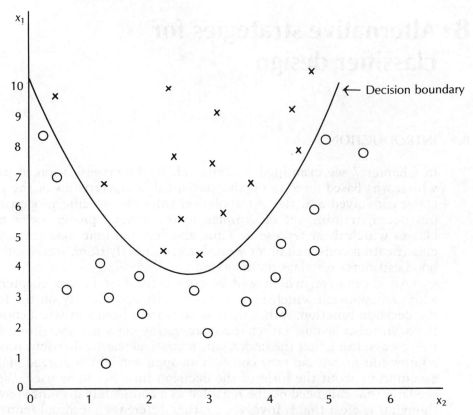

Figure 8.1 Illustration of the concept of a generalized decision surface to discriminate among pattern classes. × = class 1 points, ○ = class 2 points.

is satisfied for any pattern in feature space that lies exactly on this boundary, while for any point in feature space in the cross-hatched area $d(\mathbf{X}) > 0$ and, correspondingly, for any pattern in the double cross-hatched area of feature space, $d(\mathbf{X}) < 0$.

Hence, even without the benefit of direct visual inspection of a diagram, for an unlabeled pattern $\mathbf{X} = \{x_1, x_2\}$, it is possible to compute a value for a decision function $d(\mathbf{X})$ which will allow us to assign $\mathbf{X}$ either to class C_1 or to class C_2.

Summarizing, this decision rule is:

Assign $\mathbf{X}$ to class C_i, such that

$$d(\mathbf{X}) > 0 \Rightarrow \text{class } C_1 \text{ membership}$$
$$d(\mathbf{X}) < 0 \Rightarrow \text{class } C_2 \text{ membership}$$

(In theory we should also add that if $d(\mathbf{X}) = 0$ then we are unable to

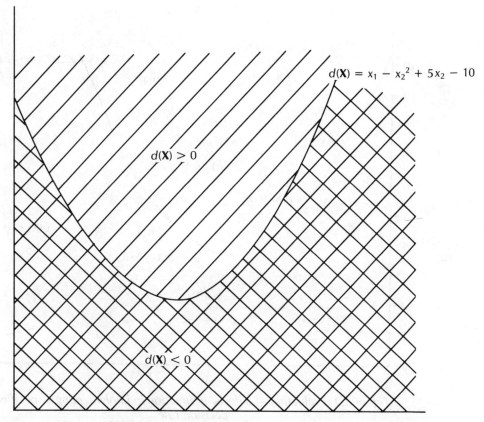

$$d(\mathbf{X}) = x_1 - x_2{}^2 + 5x_2 - 10$$

$d(\mathbf{X}) > 0$

$d(\mathbf{X}) < 0$

Figure 8.2   Relation of a geometric decision surface to a formalized decision rule.

reach a decision, but this unlikely situation can easily be resolved, either by including a $\geq$ condition, or by making an arbitrary class assignment.)

This procedure is illustrated for some simple examples in Figure 8.3. It will be clear that the success of the decision rule will depend principally on whether or not we have available enough representative samples to identify appropriate class-defining regions in feature space and whether we have chosen a good separating surface with respect to the samples available.

More important, however, is that the decision rule has been specified without making any particular assumption about the underlying form of the feature distribution functions. Of course, we have not specified how to derive the form of the required decision boundary, but since it is clear that it must always be possible in theory (even though it may be difficult to implement in practice) to specify a decision boundary to separate pattern classes in feature space (of course, the function required could in some circumstances be exceedingly complex), an approach that tackles the classification problem from this angle could be quite attractive in principle.

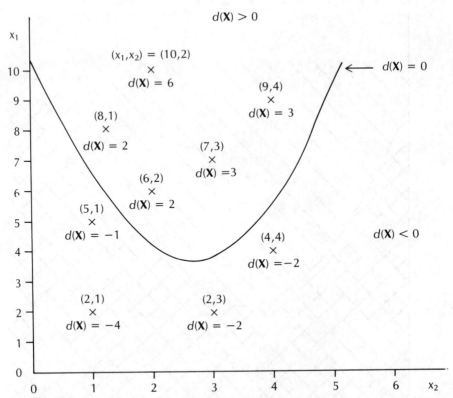

Figure 8.3 Evaluation of a decision function in implementing a specified decision rule.

This chapter will consider just such an approach where, instead of assuming something about the underlying statistical feature distributions that govern a problem domain, we instead specify a particular form for the discriminant function itself. Continuing with the example cited above, for instance, we could approach the problem by specifying a generalized quadratic function and attempt to use data samples to fit a specific set of coefficients to this generalized quadratic form. To take the completely general case, we would need to begin by assuming only that a decision function of the form

$$d(\mathbf{X}) = \varphi_1 f_1(\mathbf{X}) + \varphi_2 f_2(\mathbf{X}) + \ldots + \varphi_p f_p(\mathbf{X}) + \varphi_{p+1}$$

is to be used for classification, where $d(\mathbf{X})$ is to be evaluated by taking an appropriate number of terms ($p+1$ of them in the general case). As we have seen earlier, it can be a very difficult problem in practice to evaluate the parameters for such a decision function and consequently in practical situations we seek, as in the case of the statistical classifier, some approximations that might make for more convenient and tractable solutions, even though such approximations might introduce errors and lead to less than optimal performance. This idea was also discussed in Chapter 7.

Not surprisingly, the simplest and probably the most common approach is to work with discriminant functions which are the simplest to handle and this usually means that, where at all appropriate, we would choose to work on the assumption that a decision function has a *linear* form. As we shall see, this approach can provide an intuitively satisfying and easily manipulated method for classification.

8.2 LINEAR DECISION FUNCTIONS

The most general form of a linear decision function may be represented as

$$d(\mathbf{X}) = \mathbf{\Phi}'\mathbf{X}$$

where

$$\mathbf{X} = \{x_1, x_2, \ldots, x_n, 1\}$$

and is known as the *augmented* pattern vector* since, for mathematical convenience, the 1 has been added as an $(n + 1)$th term; and

$$\mathbf{\Phi} = \{\varphi_1, \varphi_2, \ldots, \varphi_n, \varphi_o\}$$

is a *weight vector*, the values φ_k $(k = 0, n)$ defining the coefficients of a linear equation in the variables x_k.

The approach to be considered, therefore, is to make the assumption that a decision function with the general form specified will be appropriate for classification purposes, and then tackle the more specific problem of finding a set of coefficients $\mathbf{\Phi}$ that will be suitable for a given task.

Before addressing this latter point, which would appear to be the principal area of concern, we should not ignore a fundamental question relating to the underlying philosophy of this approach; it is easy to overlook the fact that, given the assumption of linearity of decision function, we shall need to know how to utilize this function. If we are dealing with just two classes, for example, the problem is straightforward, taking us right back to the very simple methodology introduced in Chapter 6 when we dealt with the fundamental issues in pattern recognition. The general idea is restated for completeness in Figure 8.4. Moving to more realistic situations, however, leads to a less clearly defined objective.

We shall examine just one possible configuration in detail, choosing

---

*Note that in general in this section we shall use this definition of a pattern vector – i.e. in its augmented form – for convenience, without specifically adopting a new notation.

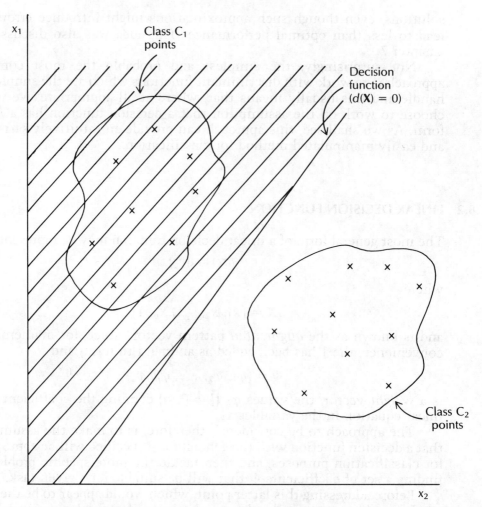

**Figure 8.4 Example of a linear decision function in the two-class case.
The cross-hatched area represents the region associated with class C₁.**

a scheme that bests fits the idea of a standard or canonical representation
of a pattern classifier design. The underlying mechanism of the proposed
classifier is best illustrated by reference to Figure 8.5. This deals with the
specific case where we have three possible pattern classes C_1, C_2 and C_3,
and where we are able to represent pattern samples in two-dimensional
feature space such that $\mathbf{X} = \{x_1, x_2\}$. This latter condition simply allows
us to visualize the ideas conveniently.

The basic idea is that we are using a set of decision surfaces
$d_{kl}(\mathbf{X})$ $[(k,l) \in \{(1,2), (2,3), (1,3)\}]$ such that $d_{kl}(\mathbf{X})$ provides a pairwise
distinction between the C_k and the C_l classes. Moreover, each decision

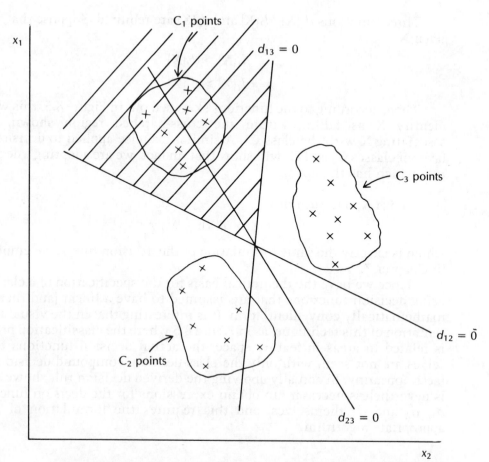

Figure 8.5 Implementation of a linear decision function for a three-class problem in two-dimensional feature space. The cross-hatched area represents the region associated with class C_1.

surface $d_{kl}(\mathbf{X})$ is derived from class-specific decision functions $d_k(\mathbf{X})$ and $d_l(\mathbf{X})$ such that

$$d_{kl}(\mathbf{X}) = d_k(\mathbf{X}) - d_l(\mathbf{X})$$

where

$$d_k(\mathbf{X}) = \Phi'_k \mathbf{X}$$

and

$$d_l(\mathbf{X}) = \Phi'_l \mathbf{X}$$

These functions can be seen to work in the following way for the three classes C_1, C_2 and C_3.

Three functions $d_1(\mathbf{X})$, $d_2(\mathbf{X})$ and $d_3(\mathbf{X})$ are required. Suppose that, for a given $\mathbf{X}$,

$$d_1(\mathbf{X}) > d_2(\mathbf{X})$$

$$d_1(\mathbf{X}) > d_3(\mathbf{X})$$

Then, according to the functions sketched out in Figure 8.5, this would identify $\mathbf{X}$ as falling within the cross-hatched region shown, thus associating $\mathbf{X}$ with the class C_1. A similar exercise applied to decisions in favor of classes C_2 and C_3 will show that, in fact, we are adopting a decision rule which has the form:

Assign $\mathbf{X}$ to C_i such that

$$d_i(\mathbf{X}) > d_j(\mathbf{X}) \qquad \forall\, j \neq i$$

which is exactly the same formulation of the decision rule as we employed in Chapter 7.

Hence we have the theoretical basis for the specification of a classifier using decision functions that are assumed to have a linear (and therefore mathematically convenient) form. It is interesting that in the visual representation of this technique, as in Figure 8.5 where the classification process is related to areas in feature space, the actual decision functions themselves are not seen, with only the class-defining compound decision surfaces appearing. In actually applying the derived decision rule, however, it is nevertheless necessary to obtain expressions for the decision functions d_1, d_2 and d_3 themselves, and this requires the formulation of some appropriate algorithm.

8.3 DETERMINATION OF LINEAR DECISION FUNCTIONS

As in Chapter 7, we shall assume that we have available a training set of sample patterns, labeled according to their *a priori* known class identity. Let us identify the jth training pattern from class C_i as $\mathbf{X}_{ij}$, and assume that $i \in \{1, 2, \ldots, N\}$.

The object of the training exercise is to find a set of weights

$$\Phi_i = \{\varphi_1, \varphi_2, \varphi_3, \ldots, \varphi_n, \varphi_o\} \qquad \forall\, i$$

such that the resulting decision functions

$$d_i(\mathbf{X}) = \Phi_i' \mathbf{X}$$

are characterized for all i classes according to the classification scheme specified in the preceding section.

There are a number of ways in which this objective may be achieved, of which one commonly considered is an algorithm that uses a fixed-increment weight vector modification procedure. This algorithm may be described in terms of the following steps:

Step 1 Assign an initial value to each of the N weight vectors Φ_i. To some extent this initial assignment of weights is arbitrary and could in a simple case involve setting $\varphi_{ij} = 0 \; \forall \; i,j$.

Step 2 Evaluate $d_i(\mathbf{X}_{ij})$ where $\mathbf{X}_{ij}$ denotes the jth training pattern of the ith class. Evaluate

$$d_k(\mathbf{X}_{ij}) \qquad \forall \; k \neq i$$

Then, (a) if $d_i(\mathbf{X}_{ij}) > d_k(\mathbf{X}_{ij}) \qquad \forall \; k \neq i$, Φ_i is not changed;
(b) for all values of k where condition (a) does not hold make

$$\Phi_k = \Phi_k - \theta \mathbf{X}_{ij}$$

$$\Phi_i = \Phi_i + \theta \mathbf{X}_{ij}$$

where θ is a constant or fixed increment.

Step 3 Repeat step 2 $\forall \; j$.

Step 4 Repeat steps 2 and 3 $\forall \; i$.

Step 5 Repeat steps 2–4 until no further change occurs in any weight vector Φ.

This algorithm will converge to a fixed set of weights provided that the training samples do in practice fall into areas of feature space that are separable by means of the linear surfaces discussed above (i.e. provided that the samples are 'linearly separable'), and subsequently unlabeled patterns may be classified by the application of the standard decision rule. In this case the performance of the classifier will depend on the extent to which the training samples were representative of the classes from which they were drawn.

Even if absolute convergence does not occur (implying that there are some samples in the training set that do not conform to the discrimination configuration in feature space already discussed) then, as with all pattern recognition problems, we may still decide that an acceptable error rate performance is possible, sacrificing some measure of performance for the undoubted benefits afforded by the resulting convenience of the linear functions and simple classifier implementation.

8.4 FURTHER CONSIDERATIONS IN CLASSIFIER SPECIFICATION

We have been dealing above with an approach to the design of a pattern classifier which assumes a given (linear) form for a decision function. Of course, this idea is not new – it is precisely what we met in Chapter 7 where design decisions rested rather on assumptions about underlying statistical distributions. For example, in that approach, when pattern classes had identical covariance matrices and features were statistically independent, it will be recalled that the form of the decision function adopted for a given class C could be reduced to:

$$d(\mathbf{X}) = \frac{1}{2\sigma^2}\,(2\mu'\mathbf{X} - \mu'\mu) + \log p(C)$$

which has a linear form (see Figure 7.8).

Indeed, it would be easy to fit this function to the 'standard' linear form developed in this chapter: by making

$$\Phi_i = \frac{1}{\sigma^2}\,\mu$$

and

$$\varphi_0 = -\frac{1}{2\sigma^2}\,\mu'\mu + \log p(C)$$

we may rewrite $d(\mathbf{X})$ as

$$d(\mathbf{X}) = \Phi'\mathbf{X} + \varphi_0$$

which is seen to be a slightly rearranged version of the generalized linear function

$$d(\mathbf{X}) = \Phi'\mathbf{X}$$

and is obtained by relabeling

$$\Phi = \{\varphi_1,\ \varphi_2,\ \ldots,\ \varphi_n\}$$

and reverting to the nonaugmented pattern vector $\mathbf{X} = \{x_1,\ x_2,\ \ldots,\ x_n\}$.

Similarly, we can extract from this discussion another basic idea which can be developed further. Recall that, in the special case above, the discussion of Chapter 7 demonstrated that a classification rule is adopted such that a test (unlabeled) pattern is assigned to that class for which its Euclidean distance from the class mean vector is a minimum. In other words, the idea was introduced that distance functions (of which Euclidean distance, as we saw, is but one possible example) might provide an approach to pattern classification. Thus, if a pattern is 'close to' or 'similar to' some specified point or area in feature space which is associated with a particular pattern class, then that pattern can be classified as itself belonging to that class. In the example quoted, a pattern class was defined in terms of its mean vector which was to be regarded as a sort of class-defining prototype. This general idea of distance-based classification, however, can be extended to give us yet another way of approaching the implementation of pattern classification algorithms. Accepting, then, that a number of issues underlying pattern classification are common to a variety of specific design approaches, we can now explore distance-based classification schemes – or clustering algorithms – a little further.

8.5 CLUSTERING TECHNIQUES AND PATTERN CLASSIFIER DESIGN

Clustering algorithms are inherently attractive in that they relate the concept of 'similarity' between patterns to the idea of proximity of patterns in feature space. This is an intuitively satisfying idea and is readily grasped. Thus, given an appropriate distance measure in some defined feature space, it is assumed that more similar patterns would be closer together in the feature space while at increasing distances patterns would be regarded as increasingly dissimilar.

Clustering algorithms are varied, but the basic principles can be illustrated readily, and we shall initially consider two related clustering schemes. However, since all clustering techniques require the use of a distance metric we shall first define the measure it is proposed to use by way of illustration. Not surprisingly, in view of the preceding discussions, the idea of the straightforward Euclidean distance measure is both a natural and a convenient choice.

Slightly generalizing the result quoted in Chapter 7, the Euclidean distance, ξ, between any two patterns represented as $\mathbf{X}_1$ and $\mathbf{X}_2$ respectively may be defined according to:

$$\xi^2 = (\mathbf{X}_1 - \mathbf{X}_2)'(\mathbf{X}_1 - \mathbf{X}_2)$$

Hence, if we map any two patterns into feature space and measure the Euclidean distance between them, the value of ξ so obtained will be interpreted as indicating the degree of similarity between them such that

$$\text{Degree of similarity} \propto 1/\xi$$

Let us consider two algorithms that might be used to demonstrate how distance measures can formally be used for pattern classification. These algorithms work on the assumption that pattern classes correspond to definable regions of feature space and that sample patterns of a given class will form a cluster in their feature space. For example, suppose that we are manufacturing high-precision components characterized by the three features area, width and perimeter. We would be able to envisage some 'archetype' or idealized sample located precisely in three-dimensional feature space. Moreover, if we are dealing with a fairly high-precision manufacturing process, we would expect relatively little deviation in any of the three critical measurements in any sample produced in the normal course of events. Such a process would generate samples that would define a particular object class as a fairly tight cluster of points in feature space (see Figure 8.6), with individual samples clustered relatively closely around the archetype or notionally ideal sample.

We shall consider two clustering procedures in turn, illustrating their application by reference to an example. For convenience, we shall choose a number of patterns represented in two-dimensional feature space, though extrapolation to higher dimensionalities, while rather less easy to visualize

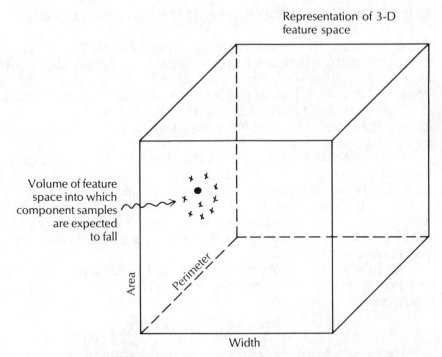

Representation of 3-D
feature space

Volume of feature
space into which
component samples
are expected
to fall

Area

Perimeter

Width

Figure 8.6 An illustration of sample clustering within a class in three-dimensional feature space. • = archetype or 'ideal' sample, × = individual component samples.

directly, cause no great difficulty conceptually. Patterns depicted in this feature space will be referenced simply by a number for subsequent identification purposes.

8.5.1   Archetype-based clustering

Let us assume that we are dealing with a three-class problem, where classes C_A, C_B and C_C may be assumed to be characterized by the archetype or class-defining patterns labeled 1, 2 and 3 respectively (see Figure 8.7).

The simplest possible clustering algorithm applicable in these circumstances is to classify an unknown pattern such as that shown as pattern 16 in Figure 8.7 by evaluating its distance from each of the archetypes in turn and, exploiting the notion of the inverse relationship between distance and similarity, assigning the unknown pattern to the class associated with the nearest archetype.

Designating as ξ_a the Euclidean distance between a pattern of interest and any pattern referenced as a, the algorithm may be summarized as in

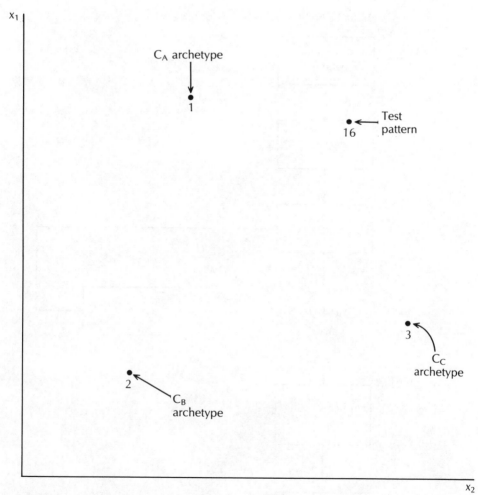

Figure 8.7 Archetype representation in feature space.

Figure 8.8, and its application (using 1 cm = 1 unit of distance) yields the following distances of interest with respect to test pattern 16:

$$\xi_1 = 4.2$$

$$\xi_2 = 8.8$$

$$\xi_3 = 5.5$$

The unlabeled pattern is consequently assigned to class C_A. It may be seen that, in principle, the underlying mechanism of this classifier parallels very closely that of the minimum distance classifier discussed in Chapter 7, where the class archetype was defined to be the vector of means of the feature measurements.

Assume: Archetypes available for classes $C_1 \ldots C_N$

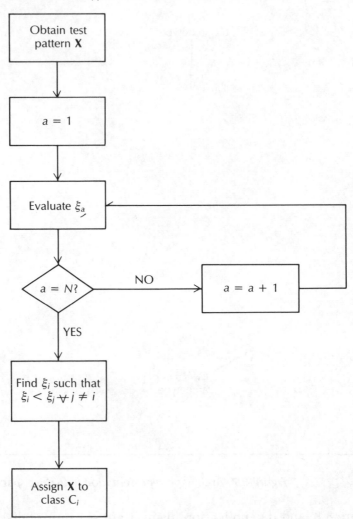

Figure 8.8 Basis of an algorithm for an archetype-directed clustering procedure.

8.5.2 Generalized nearest neighbor clustering

The preceding somewhat naive algorithm may be inappropriate when good class-defining archetypes cannot be identified or when clusters are less compactly formed. A rather more flexible variant of the clustering approach, known as the k-nearest neighbor or k-NN procedure, offers an alternative in these circumstances.

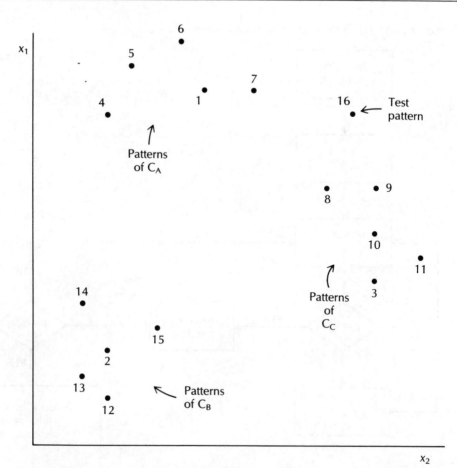

Figure 8.9 An illustrative set of pattern points in two-dimensional feature space.

The idea of this algorithm is to use a set of known-identity pattern samples and, for classification of an unknown sample, invoke a rule that uses a majority decision in selecting the set of patterns in a class-cluster to which the unknown sample is most similar. Again, this approach may be easily illustrated, and is made clear by reference to Figure 8.9. This shows three pattern classes formed from labeled examples – C_A (comprising patterns 1,4,5,6 and 7), C_B(2,12,13,14 and 15) and C_C(3,8,9,10 and 11). In order to classify the unidentified pattern (16) its distance from all labeled samples is measured. For application of the k-nearest neighbor procedure, we then select the k samples for which the measured distance is smallest, and assign the unidentified pattern to that class whose identity label occurs most frequently among these k measurements.

The algorithm is formally illustrated in Figure 8.10, and yields the

Let: $\mathbf{X}_{i\alpha_i}(i=1,N)$ be the α_i available samples from class C_i

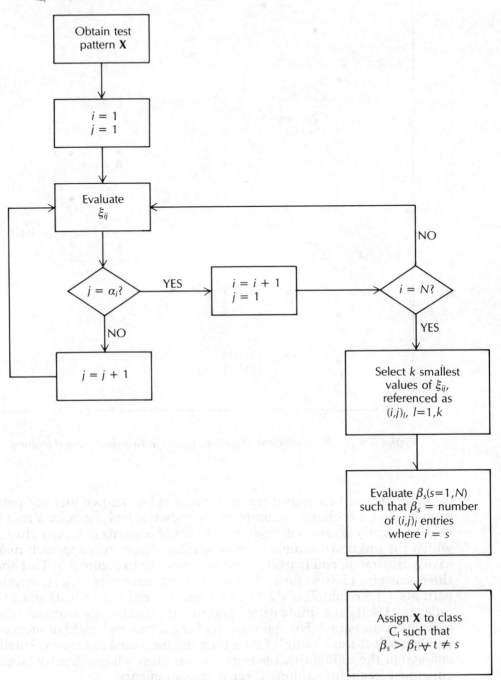

Figure 8.10 Basis of an algorithm for a k-NN clustering procedure.

Table 8.1 Relevant parameters in the application of a k-NN clustering procedure for the test data shown in Figure 8.9

Class	Pattern reference	ξ = distance from test pattern **X** (16)
A	1	3.9
	4	6.4
	5	5.9
	6	4.9
	7	2.6
C	3	4.4
	8	2.0
	9	2.0
	10	3.2
	11	4.2
B	2	9.0
	12	9.9
	13	9.9
	14	8.2
	15	7.1

Decision procedure

k-nearest neighbors:

Pattern reference	8	9	7	$\Big\} k = 3$
Associated class	C	C	A	

Assign to
class C_C

Pattern reference	8	9	7	10	1	$\Big\} k = 5$
Associated class	C	C	A	C	A	

Assign to
class C_C

results shown in Table 8.1 which arbitrarily uses a 5-NN rule and assigns the unknown pattern to class C_C. (Note that here also a 3-NN procedure would have had a similar result.) Visual inspection should confirm that, assuming the features adopted to be reasonable in indicating class membership, this represents a rather more intuitively satisfactory result than that obtained with the original algorithm. This is principally because, in the application of this algorithm, we have taken more account of the overall class-defining properties of a group of samples within a class, and this has revealed a possibly inappropriate choice of archetype in the earlier case.

Of course, these two examples do not exhaust the possible variations, but they do illustrate the general approach. Furthermore, a rather different perspective on pattern classification may be obtained by considering the general philosophy of clustering techniques in an alternative way, as the following section will demonstrate.

8.6   SELF-REGULATING CLUSTER FORMATION

In all the pattern recognition tasks considered up to this point we have consistently made the assumption that known identity samples – a training set of patterns – would be available. In cases where this is not possible, however, provided that pattern distributions are such that within a class patterns are similar (in a clustering sense) with respect to the chosen features, and that patterns of different classes have a tendency to be dissimilar, we find that clustering techniques can in principle provide a way of actually *identifying* pattern classes in addition to enabling the subsequent identification of unlabeled samples.

The underlying idea is that a pattern of interest is examined in relation to others that have been encountered and is added on to an existing cluster if it is deemed to be sufficiently similar to its existing members. A pattern that falls outside an arbitrarily chosen distance from any existing cluster then forms a new cluster which can itself attract further members as more examples are seen, and so on. Figure 8.11 shows a simplified version of such an algorithm which employs (arbitrarily defined) archetypes to define cluster centers, in rather the same way as the procedure outlined in Section 8.5.1 operated.

Again, an example will serve to illustrate this process. If the 16 patterns already encountered to previous examples are taken in the randomly chosen order:

$$1, 2, 6, 10, 4, 12, 11, 5, 14, 9, 13, 8, 15, 7, 3, 16$$

then the application of the algorithm proceeds as shown in Table 8.2 where a threshold $\xi_T = 3.3$ has been chosen. The result is to cluster the patterns such that three classes are identified, with individual pattern samples clustered as shown in Figure 8.12. It can be seen that this has resulted in the classification we started with in the preceding case, with the previously unlabeled pattern referenced as 16 assigned as it was according to the 5-NN rule. It will be noted, however, that the arbitrary ordering of the patterns to be processed has led to pattern 10 being adopted as the archetype for class C_C in this case.

Although the algorithm has worked well in this example it should be apparent that the actual results produced will depend to a large extent on

Let $\mathbf{X}_i$ = pattern representing ith class cluster
Let C_i = class defined by $\mathbf{X}_i$

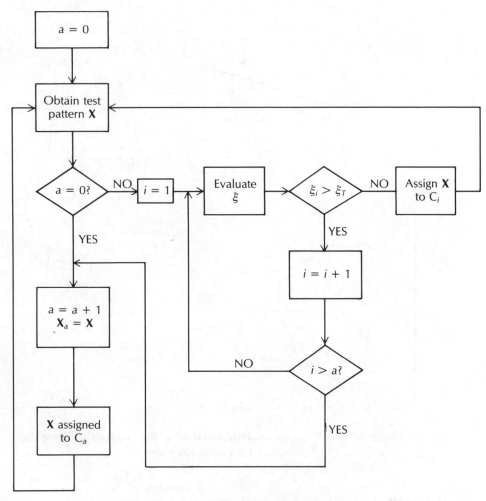

Figure 8.11 Basis of an algorithm for self-regulating cluster determination.

variable factors such as the threshold ξ_T chosen, and once again variations on this basic scheme can give significant improvements. It should be a simple matter, for example, to modify the scheme so that cluster assignments can be made according to a form of the k-NN rule. Similarly, a good deal of trial and error, perhaps involving progressive merging of identified clusters, may be necessary if optimal levels of performance are to be attained.

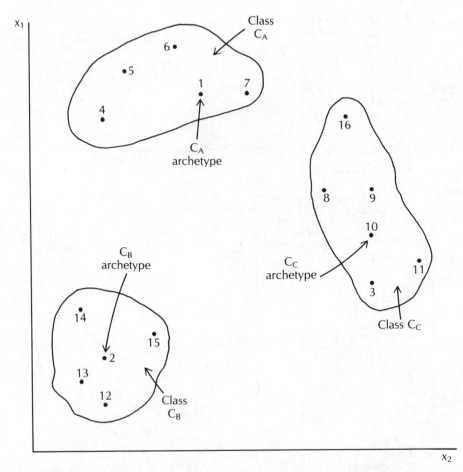

Figure 8.12 Feature space representation of the result of applying the cluster-locating algorithm.

8.7 SUMMARY

This chapter has shown how, building on an underlying statistical theory of pattern recognition, alternative strategies for classification may be adopted which can be effective in differing circumstances. As we saw in Chapter 6, it is important to the designer of a pattern recognition system that a range of tools is available with which to tackle a particular problem. The ultimate choice of technique will then be made on the basis of the available data, the extent to which valid assumptions about the pattern-generating environment can be made, relative computational requirements, and so on.

***Table 8.2 Application of a cluster-locating algorithm to patterns in
the defined sample set, taking threshold*** $\xi_T = 3.3$

Input pattern	Class A, cluster center 1	Class B, cluster center 2	Class C, cluster center 10
1	—/A*	NS	NS
2	11.7/B*	NS	NS
6	2.2/A	—	NS
10	9.2/—	12.1/C*	NS
4	4.0/A	—	—
12	13.6/—	2.0/B	—
11	11.4/—	13.2/—	2.2/C
5	3.2/A	—	—
14	10.3/—	2.2/B	—
9	8.1/—	13.0/—	1.9/C
13	13.0/—	1.4/B	—
8	6.4/—	11.4/—	2.9/C
15	10.2/—	2.3/B	—
7	2.0/A	—	—
3	10.6/—	11.4/—	2.0/C
16	6.0/—	14.2/—	5.0/C

Key:
X/Y = distance/class assignment
*denotes creation of new cluster
NS = not started

Class-cluster formation
          Constituent patterns
Class C_A 1, 6, 4, 5, 7
Class C_B 2, 12, 14, 13, 15
Class C_C 10, 11, 9, 8, 3, 16

Perhaps most important of all, it should now be becoming clearer that the techniques introduced are all fairly closely related, reflecting at the lowest level the fact that classification is basically a statistical process, where concepts such as linearity in discriminant functions, minimum distance procedures, and so on, are merely practical expressions of the search for tractability in implementation within a generalized theory and methodology for the classification process.

In this context it is even possible to approach the pattern recognition problem from a point of view that de-emphasizes the calculation of statistical properties or the specification of geometrical form in feature space of the decision functions employed, instead concentrating on a computational structure that is relatively simple to engineer practically. The next chapter will introduce some ideas on pattern classification from just this viewpoint.

8.8   EXAMPLES FOR SELF-ASSESSMENT

8.1   Show how it is possible to derive a classifier configuration that utilizes p linear discriminant functions, each discriminating between one specific pattern class and all other classes, to obtain a classification decision among p possible classes.

Show how the scheme may be represented in two-dimensional feature space and hence comment on the possible drawbacks of this configuration.

8.2   For the data shown in Table 8.3:

(a)   Assuming that pattern numbers 1, 18 and 25 are to be regarded as class archetypes, use a clustering technique to determine the identity of the remaining patterns.

(b)   Taking the patterns in ascending order of their numerical labels, use a cluster-seeking method to evolve a possible set of class identities.

Table 8.3

Pattern reference number	Descriptor value		
	x_1	x_2	x_3
1	47	12	8
2	41	12	12
3	11	48	9
4	9	29	33
5	10	39	12
6	36	11	14
7	5	42	46
8	11	39	38
9	18	45	17
10	28	26	31
11	34	18	11
12	7	41	44
13	44	15	9
14	46	10	6
15	9	38	12
16	14	28	24
17	9	41	9
18	6	45	41
19	8	42	6
20	17	40	29
21	40	21	16
22	42	14	7
23	12	38	48
24	14	28	34
25	8	44	11

Table 8.4 Measurements on known samples (arbitrary units)

Normal		Abnormal	
L	B	L	B
1.4	0.8	2.0	0.6
1.0	0.2	2.6	0.8
1.0	0.4	2.2	1.4
1.0	0.8	2.4	2.0
0.8	1.4	1.8	1.6
0.6	0.4	1.6	2.2
0.6	1.0	1.4	1.8
0.2	0.6	0.4	2.0

How do the clusters differ as different cluster distance thresholds are adopted?

8.3 A clinical screening process requires visual imaging of a tissue sample treated such that an area of interest is highlighted by dark staining. Measurements on previously obtained samples have been shown to be indicative of whether a sample is normal or abnormal (see Table 8.4), where L and B refer respectively to the average length and breadth of the highlighted area, consistently measured in relation to some specified alignment of the sample.

Use this information to derive a decision function and hence specify a decision rule that allows categorization of further samples.

Explain how this decision rule might be modified if:

(a) an alternative, highly reliable, though very time-consuming and expensive, testing procedure is known;
(b) the consequences of failing to identify an abnormality could be very serious.

Briefly indicate what processing operations need to be carried out on the images to enable the derivation of the required measurements and comment on how these might influence the classification algorithm.

9 ★ Memory network approach to pattern recognition

9.1 INTRODUCTION

This section of the book has been concerned with the ways in which algorithms for pattern recognition might be developed and implemented. We began with a consideration of the underlying mechanisms of pattern recognition; and consequently developed the basis for appropriate algorithms to carry out classification tasks; and then saw how, operating within the framework established, it is possible to derive alternative approaches to classification which can be more readily applied to a variety of problem domains.

To complete this section we shall turn our attention to yet another approach to pattern recognition, and one that differs from the others we have considered in that, from one point of view, it might be regarded as having its roots more in the realm of engineering than that specifically of mathematics. This statement needs some qualification – it is not intended to imply that the technique has no appropriate or useful formulation in mathematics, or that its only usefulness is as a rigidly defined and physically constructed system. It is rather that the technique to be described is most easily visualized and understood when seen as a *physical system* rather than an algorithm or a mathematical process.

It will be apparent that, as is inevitably the case, the operation of the method is dependent ultimately on the underlying statistics of the pattern environment in question, but the technique has some considerable advantages, including the following:

1. It is conceptually very simple.
2. It utilizes pattern features that are directly and conveniently available.
3. It affords great flexibility in terms of its possible means of implementation.
4. Its parallel or semi-parallel structure offers the potential for very high operating speeds.
5. It is very easily applied in practice, because the information required to compute decision functions is obtained from an on-line 'self-adaptive' procedure.

As we shall see, in common with all methods of pattern recognition, it has some restrictions on its range of applicability, and system performance will depend on the particular problem domain of interest. One particular constraint which is of some immediate importance is that in general it is necessary that data to be processed is in binary form, although this is not necessarily a particularly restrictive condition in the present context. The technique in question is based on a version of a general group of techniques known as *n-tuple techniques* or sometimes '*memory network classification structures*, and where the features which are used to compute a classification decision are groups or *n-tuples* of (binary) pixel points. Such methods have a relatively long history in the pattern recognition field, but have in recent years gained an enhanced degree of commercial viability.

We shall first of all describe the technique in its simplest form and in purely conceptual terms. Subsequently we will establish its relation to other algorithms we have considered, make some comments about issues relating to implementation, and finally discuss some possible extensions to the technique.

9.2   CONCEPTUAL BASIS FOR MEMORY-NETWORK CLASSIFIERS

We shall take a small-scale example in order to define terms and establish principles, generalizing this later. We shall assume that a pattern may be defined as $\mathbf{X} = \{x_1, x_2, \ldots, x_k\}$ where $x_i \in \{0,1\} \, \forall i$. While in principle the feature elements x_i may be any two-valued features, it is most common to regard them as directly obtained binarized pixel values when two-dimensional visual images are being considered. We may then begin with two simple definitions, as follows.

> An *n*-tuple is a subset of the elements of $\mathbf{X}$ which are regarded as an ordered group. For example,
>
> $$i\text{th } n\text{-tuple} = \{x_p, x_q, \ldots, x_r\}$$
>
> where $p, q, r \in \{1, \ldots, k\}$.

> The *state* of an *n*-tuple is defined by the (binary) values assigned to the elements of the *n*-tuple at some particular time. For example,
>
> $$n\text{-tuple state} = \{1, 1, 0, \ldots\}$$

Figure 9.1 illustrates this for a small-scale low-resolution two-dimensional image matrix, Figure 9.1(a) identifying an arbitrarily chosen *n*-tuple (actually a 4-tuple in this case), and Figure 9.1(b) showing the state of this particular *n*-tuple when some specific image data is under consideration.

The basic mechanism of the memory-based classifier scheme is to work with *n*-tuple features, storing information about the occurrence or nonoccurrence of particular *n*-tuple states in a set of *training patterns*

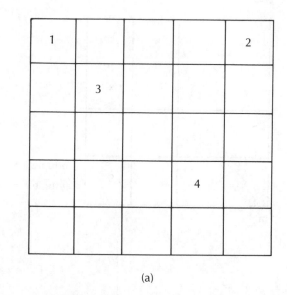

(a)

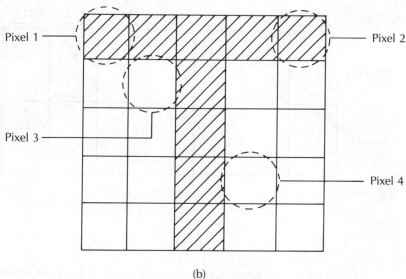

(b)

**Figure 9.1   Identification of n-tuples and n-tuple states. (a) n-tuple
(4-tuple) formed by order pixel group {1,2,3,4}, (b) state of n-tuple
{1,2,3,4} is {1100}.**

which are drawn from pattern classes of interest. More precisely, a pattern
X of unknown identity is classified by such a system according to the
pattern class which generates the maximum number of n-tuples in **X** with
a state corresponding to those occurring at least once in the patterns of
the appropriate training set. This is made much clearer if an example is

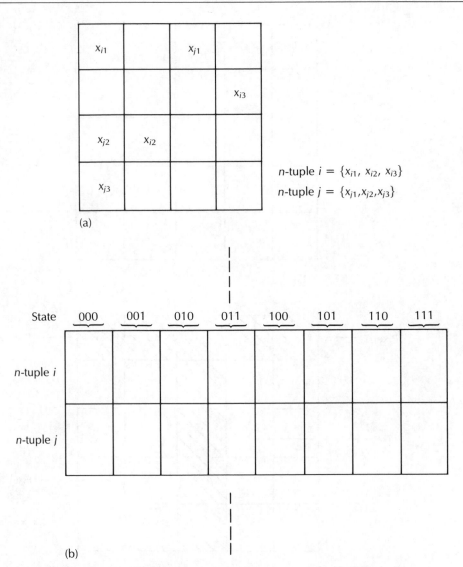

Figure 9.2 *(a) Identification of reference 3-tuples for illustrative purposes. (b) Memory array required to identify possible states occurring in training samples.*

considered, and we will for the moment keep to the small-scale format for convenience and clarity. Figure 9.2 shows a 4 × 4 image matrix with two arbitrary n-tuples (3-tuples, in fact) identified for illustration. As we shall see shortly, we are going to record information about the state of these n-tuples, and consequently Figure 9.2(b) shows a table which lists all the possible states that may be associated with all possible n-tuples. Since we are working with 3-tuples in the example, each such feature may be asso-

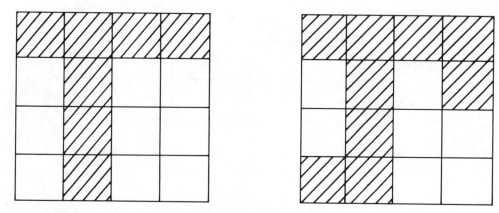

Figure 9.3 Pattern samples from a class T.

ciated with any one of eight possible states (in general, 2^n states). The dotted lines merely serve to indicate that in practice many more n-tuples might generally be considered. The idea of a 'memory-based' classifier is related to this very aspect of the system, namely that n-tuple state occurrences are stored in this table, but we shall return to this point in due course.

The operation of a classifier system based on these principles is divided into two stages or phases. Clearly, it is necessary to *train* the system to respond to a given set of pattern classes – this constitutes the *training phase* – while subsequently, in a recognition or *processing phase*, the system is used to classify a pattern **X** of unknown identity. We will consider each of these phases, beginning with the training phase.

Let us first deal with a system required only to perform a dichotomization process. In other words, classification consists in deciding whether **X** belongs to a chosen class of interest or whether it does not. Figure 9.3 shows two examples which might be available of digitized samples of the alphabetic character T. Provided that we know *a priori* the true identity of these patterns in this way, they may be used to train the system to respond to this particular pattern class.

This is illustrated in Figure 9.4. Training consists in exposing the system to each training pattern in turn and 'flagging' the occurrence of particular n-tuple states in each case. For the first training pattern the ith n-tuple assumes the state 110 and the jth 100. These states are recorded by storing a 1 flag in the appropriate 'addresses' of the stored table. A similar procedure records n-tuple states occurring in the second training pattern, as Figure 9.4 also shows.

Thus, at the end of the training procedure as defined above, the stored table contains an implicit description of the pattern category from which specific training exemplars were drawn. This 'description' is somewhat abstract, of course, but is clearly related to the distribution of possible

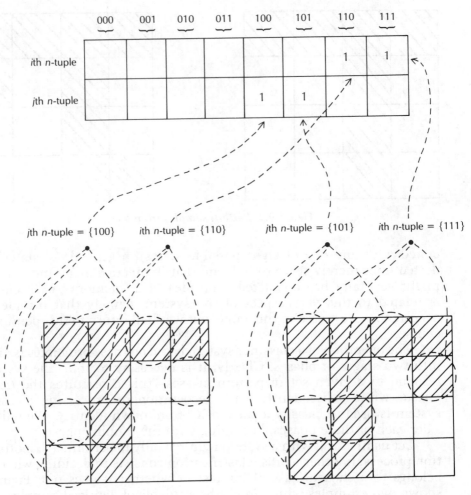

Figure 9.4 n-tuple state formation in a training operation.

'features' derived from pixel groupings occurring in an assumed representative sub-set of patterns in a particular pattern class (class T in this case). Once this phase has been completed, i.e. a representative selection of patterns has been shown to the network, this stored information may subsequently be used as the basis for a classification algorithm for patterns of unknown identity. In particular, the technique works on the assumption that the likelihood that an unknown pattern **X** belongs to the same category as the patterns in the training set is related to the occurrence of common features in **X** and the composite set of training patterns.

This can be seen more clearly by now considering the 'response' of the system to the two unidentified patterns shown in Figure 9.5. Figure 9.5(a) shows an example of a pattern which is clearly from the T-class (though

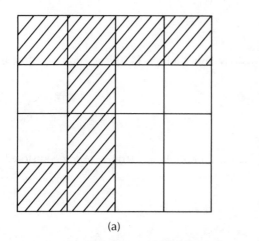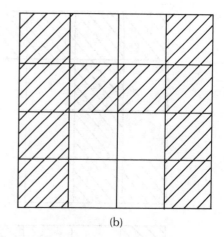

(a)   (b)

**Figure 9.5 Definition of some sample test patterns: (a) class T,
(b) class H.**

not identical to any of the training patterns), while the pattern in Figure
9.5(b) is clearly from a different category (in this case a different alphabetic
character, H). In determining system response, the n-tuple states form
addresses which access the information stored during the training phase.
The occurrence of common n-tuple states in **X** and any training pattern
generates the address of a stored 1, while previously unencountered n-tuple
states generate an address pointing to the storage of a 0. A simple arith-
metic summation over the set of n-tuples chosen then gives a convenient
measure of the likelihood that **X** belongs to the class of the previously
encountered training patterns. This is illustrated in Figure 9.6, which
shows that, with respect to the two n-tuples selected for our illustrative
example, the T-pattern generates a high (maximum) response of 2 (i.e. both
n-tuple states in the test pattern have occurred in at least one training
pattern), while the H-pattern, a member of a class other than the training
class, generates a low response of 0.

  This small-scale example demonstrates the conceptual basis of the
technique, and these ideas are readily extended to more practical structures. First of all, of course, it is more practicable to consider many more n-
tuples and indeed, if an $N \times N$ pixel image is sampled with an n-tuple
structure in such a way that each pixel is sampled exactly once, the resulting N^2/n stored n-tuple state maps represent a typical upper bound on the
requirements of the system. The resulting stored array constitutes a
memory network.

  Secondly, a system based on the structure described is able only to
distinguish between one class (the trained class) and all other possible
classes. Thus, to realize a conventional classification scheme, it is necessary to adopt one such network for every class of interest, with each individual network being trained on patterns from only its own class. For

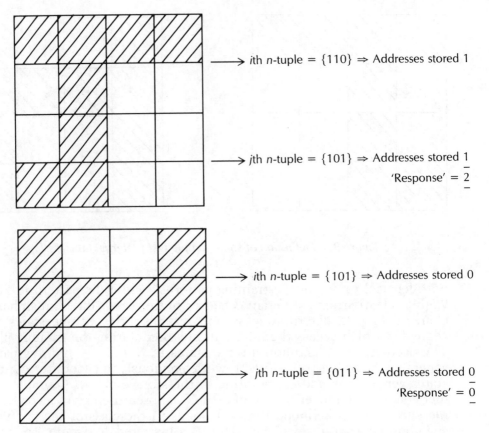

$\longrightarrow$ *i*th *n*-tuple = {110} ⇒ Addresses stored 1

$\longrightarrow$ *j*th *n*-tuple = {101} ⇒ Addresses stored 1

'Response' = $\underline{2}$

$\longrightarrow$ *i*th *n*-tuple = {101} ⇒ Addresses stored 0

$\longrightarrow$ *j*th *n*-tuple = {011} ⇒ Addresses stored 0

'Response' = $\underline{0}$

Figure 9.6 Evaluation of system response for designated test patterns.

classification, an unidentified pattern **X** is presented simultaneously to each trained network. Each responds with a 'likelihood' measure (the network numerical response defined above), and the network generating the maximum response is deemed to identify the class to which **X** should be assigned. This multicategory classifier structure is illustrated conceptually in Figure 9.7. It is apparent that each network response may be considered as a class-based decision function $f(\mathbf{X})$, and thus the operation of the memory-network classifier may be summarized formally in the following way.

Each possible recognition class C_k is associated with a processing network or memory cell array, where the *i*th cell processes a pattern sample denoted by

$$\mathbf{X}_i^{(k)} = \{x_{i,1}^{(k)}, x_{i,2}^{(k)}, \ldots, x_{i,n}^{(k)}\}$$

where $\mathbf{X}_i^{(k)}$ is a subset of an input pattern **X**.

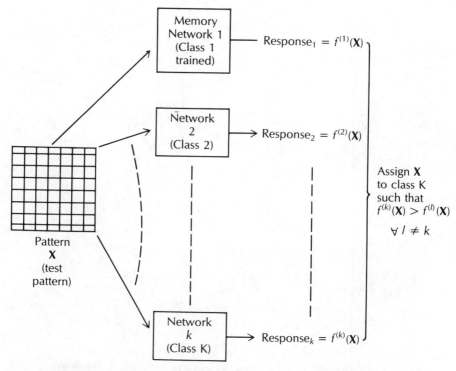

Figure 9.7 *Conceptual structure of a multi-class classification configuration.*

The C_k network is trained with patterns from class C_k, the ith memory cell storing a 1 at an address defined by the sample $\mathbf{X}_i^{(k)}$. In the subsequent classification of an unidentified pattern $\mathbf{X}$, the kth network computes a decision function $f^k(\mathbf{X})$ corresponding to the numerical summation of binary cell outputs, each signifying whether the sample $\mathbf{X}_i^{(k)}$ has or has not occurred in at least one training pattern from C_k. The unknown pattern $\mathbf{X}$ is classified as a member of C_k such that

$$f^k(\mathbf{X}) > f^l(\mathbf{X}) \qquad \forall \, l \neq k$$

This is seen to accord exactly with the classification rules considered in the preceding chapters. Similarly, it is clear that the idea of a confidence level or rejection margin can easily be introduced in just the same way as described previously. This brings corresponding benefits in terms of flexibility and the potential for trade-off between error rates and rejected or unclassified test patterns. The effect of varying a rejection margin corresponding to unit changes in the value of the decision functions adopted is illustrated in Figure 9.8, where results are based on an experiment in the classification of alphabetic and numeric characters.

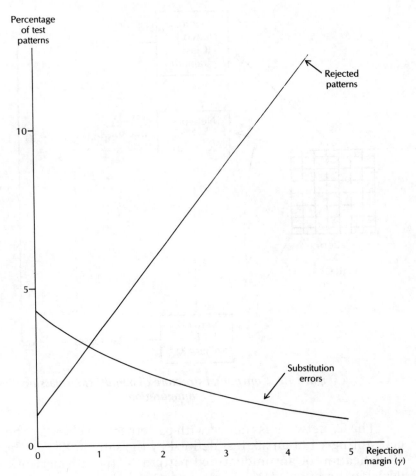

Figure 9.8 Illustration of the effect of introducing a rejection class to the classification algorithm.

9.3 CHARACTERISTICS OF MEMORY-NETWORK CLASSIFIERS

It is important to consider two particular areas in a little more detail. The first concerns the actual parameters that define a particular memory-network configuration. It should be apparent from the preceding discussion that two principal factors – under the direct control of the system designer – are fundamental in determining the actual performance of a particular system. These are:

1. the number of training patterns used in determining the contents of the stored feature matrix, and
2. the size of the *n*-tuple selected, i.e. how many pixels are grouped together in defining 'features'.

The essence of training the system is that a *representative* sample of class-defining patterns should be used in generating the stored memory functions. As a general rule, if too few training patterns are used, the system will not build up an accurate representation of the range of n-tuple states that can legitimately arise within a pattern class, while if too many training patterns are employed an increasingly large number of relatively infrequently occurring (i.e. atypical) n-tuple states will be flagged, giving rise to a saturation phenomenon, whereby network response tends to be very high whatever the origin of a test pattern **X**. Consequently, other things being equal, for a typical set of patterns the performance (measured, for example, as the percentage of test patterns correctly identified) of a system will frequently follow the sort of variation shown in Figure 9.9 with respect to a varying amount of training exposure. For best performance, therefore, it is important to choose the optimum amount of training exposure.

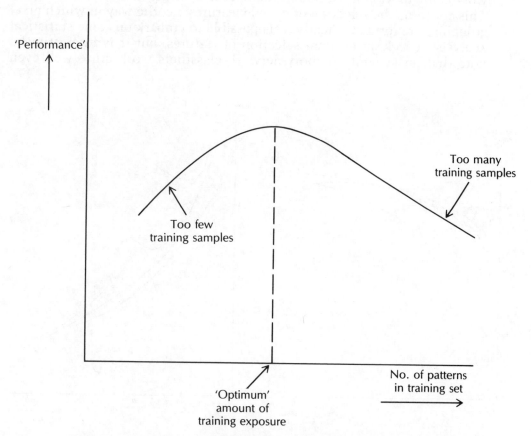

Figure 9.9 *Typical expected performance as a function of the amount of exposure to class samples in a training procedure.*

The other factor, the choice of n-tuple size, is also very important in determining performance. There are two interesting effects here. On the one hand, as n increases the system effectively takes an increasing account of the statistical interdependence of pixel features (that is, effectively, the implicit estimate of class-conditional probability distributions improves), giving a generally improving performance. On the other hand, one has to pay for this by increasing the number of training samples employed so that the probability of occurrence of the rapidly increasing number of possible different n-tuple states is maintained at a reasonable level. For the same reason, increasing n-tuple size increases dramatically the amount of storage capacity required to store the class-defining binary arrays. These points are summarized in Figure 9.10, which shows some results of a typical classification experiment, where the data used was drawn from a complete set of alphabetic characters and the ten numerals.

In passing it should also be said that another structural feature of network configuration must be specified in some way by the system designer. This concerns the selection of n-tuple features, i.e. the way in which pixel groupings are formed. Clearly, it is possible to embark on some statistical analysis to seek an optimal selection of features, but it is interesting to note that, in general, memory-network classifiers work quite well even

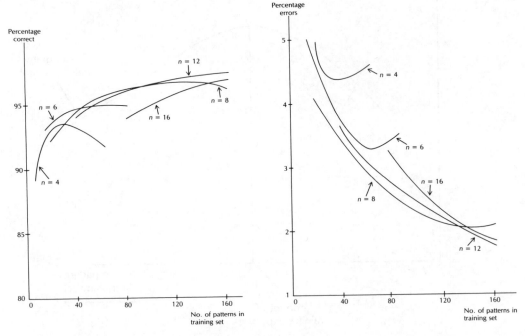

Figure 9.10 Some performance curves obtained in an experiment on alphanumeric character recognition (involving 34 pattern classes).

if no formalized selection algorithm is adopted, and where instead the designer relies on the random selection of pixel groupings to form n-tuples.

9.4 IMPLEMENTATION OF MEMORY-NETWORK CLASSIFIERS

Section 9.3 has introduced the conceptual form of the memory-network classifier. However, one of the interesting and potentially very attractive features of the technique is its amenability to implementation in a variety of forms, and it is useful to consider this aspect of classifier specification further.

First of all it is helpful to consider the structure of such a classifier in an alternative way, and it is convenient to visualize the system as shown in Figure 9.11.

A network is seen as a collection of computational *cells*, each cell computing a Boolean logic function. More succinctly, an n-tuple state is seen as forming the n-bit address to a $2^n \times 1$ memory element which stores a logical 1 (via a clocked data input channel) in each address generated in any available training pattern. In subsequent use, the memory output

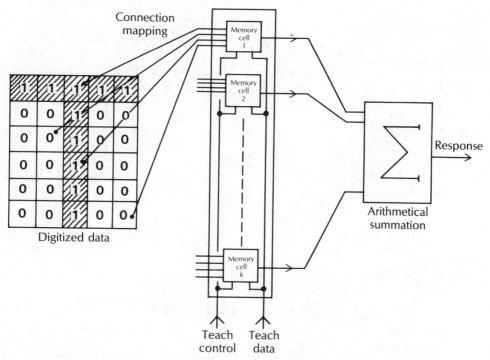

Figure 9.11 *Internal organization of a single processing network.*

channel reads the data stored in the address specified by the n-tuple state in a test pattern, and it is the summation of these output signals for all cells in a network that defines the network decision function. Thus, a multi-category pattern classifier consists of a set of such networks (see Figure 9.12), each computing a class-conditional decision function on the basis of which an overall classification decision may be made by choosing a maximum value.

This scheme is seen to offer the flexibility for a significant degree of trade-off between implementation requirements, processing speeds and operational convenience. At the lowest level the system is structurally highly parallel (the n-tuple samples could be processed simultaneously), while a more 'efficient' utilization of conventional random access memory storage (see, for example, Figure 9.13), could reduce implementation com-plexity while decreasing the intrinsic degree of parallelism in the processing structure. Finally, of course, at the highest level, the system may be simu-lated directly in a single conventional serial processor, giving significantly slower processing speeds but exploiting minimally complex hardware.

Indeed, various other schemes based on these principles are possible, including configurations that use decision functions based on the frequency of occurrence of n-tuple features, those that minimize storage require-ments by adopting alternative n-tuple configuration strategies, and those

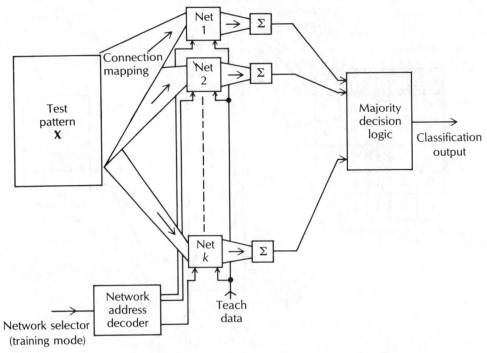

Figure 9.12 Design structure of a multicategory classifier.

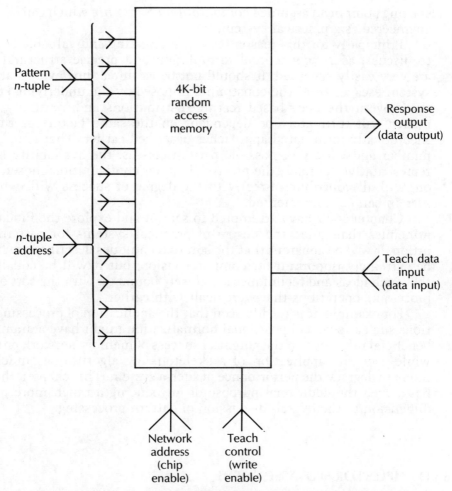

Figure 9.13 Basis of a possible implementation configuration for an n-tuple classification system.

that improve recognition performance by successively refining classification decisions in multilayer structures.

9.5 SUMMARY

In this chapter we have investigated a further approach to the implementation of an algorithm/technique for application to pattern recognition problems. The way in which this technique has been described and presented is such that it approaches the business of classification very much from the starting point of an assumed *computational structure* which can be directly engineered as a practical system.

Principally for this reason it is an interesting and valuable technique to discuss, as its conceptual formulation and precise structural details are very easily specified. It should not be assumed, however, that such a system escapes from the constraints imposed by the underlying statistics that govern the class-based feature distributions and, naturally, performance will ultimately be dependent on the same factors as the more theoretically formulated approaches described earlier. That is to say, the number and variety of possible pattern classes, the availability of class-representative samples, the precise classifier configuration chosen, and so on, will all contribute directly to the degree of success with which the system can be implemented.

Chapters 6–9 have attempted to set out and explore the fundamental principles that guide the design of practical systems for pattern recognition. It will be apparent that the sort of techniques described are likely to be extremely important for computer vision, but it will be equally clear that such ideas and techniques are closely bound up with the sort of image processing operations that were dealt with earlier.

For example, it is readily seen that the application of processing operations such as size and positional normalization might have a dramatically beneficial influence on the potential success of memory network processors while, say, the application of a skeletonizing algorithm is much more likely to degrade the performance of such a system. This chapter, therefore, has served the additional purpose of introducing a much more practical dimension to the overall discussion of pattern processing.

9.6 EXAMPLES FOR SELF-ASSESSMENT

9.1 Consider a group of just three memory cells forming an *n*-tuple network, connected to the points of a 3×3 binarized matrix as shown in Figure 9.14. The stores of the cells are initially cleared (i.e. set to logical zero) and in training the network is trained to respond to an input pattern by generating a logical one at all three outputs.

Referring to the set of input patterns shown in Table 9.1:

(a) Train the network on P_1 and determine the response to P_1, P_2, P_3, P_4 and P_5.

(b) Train the network on P_1 and P_2 and determine the response to P_1, P_2, P_3, P_4 and P_5.

(c) Change the training scheme so that P_1 is taught to produce $Z_1 = Z_2 = Z_3 = 1$, and P_6 to produce $Z_1 = Z_2 = Z_3 = 0$. Determine the response to P_1 and P_2.

Using these small-scale results as illustrations, discuss the performance

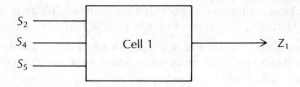

Input matrix reference

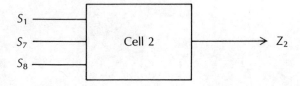

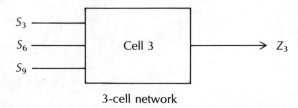

3-cell network

Figure 9.14

Table 9.1

Pattern reference	S_1	S_2	S_3	S_4	S_5	S_6	S_7	S_8	S_9
P_1	1	1	1	0	1	0	0	1	0
P_2	0	0	1	1	1	1	0	0	1
P_3	0	1	1	0	1	0	0	1	0
P_4	1	0	1	1	1	1	1	0	1
P_5	0	1	1	0	1	1	0	0	1
P_6	1	0	1	1	1	1	0	0	1

of the network as a simple pattern classification system for this particular set of patterns.

9.2   Explain how the system structure decribed in the text might be modified to weight the contribution of individual cells to the overall network response according to the frequency with which its current n-tuple state had occurred during training exposure.

   How would you expect this mechanism to change the performance characteristics of this type of classifier and why?

9.3   Discuss the viability of applying an n-tuple recognition system to a task involving the recognition of images that have been processed using a skeletonization algorithm.

10 ★ A postscript

The automatic processing of images, the automatic interpretation of visually sensed data and the development of computer vision algorithms are areas which computer scientists and engineers of the future are unlikely to be able to ignore. Such is the scope for the potential application of computer vision systems that it would be very difficult to produce an exhaustive or adequately comprehensive list. However, such a list would almost certainly include, by common consent, the following application areas of immediate significance:

Manufacturing systems
Automatic inspection/quality control
Operation in hazardous environments
Medical screening and diagnostic aids
Design of better man–machine interfaces
Automatic warehousing
Vehicle guidance systems
Office automation/document processing
Design of robotic systems
Surveillance systems
Analysis of remotely sensed data

and the reader will not find it difficult to extend and amplify this initial list.

Such diversity in possible application demands the formalization of basic concepts and techniques that can be utilized/modified/refined in a variety of ways to suit specific problems and differing task domains, but it is important that the underlying issues are clear and that the elemental techniques from which more complex processes can evolve are mastered. This book has endeavored to address some of these basic issues, to prepare the reader to extend and refine his knowledge from a sound initial base.

The framework for the particular techniques introduced in the main body of the text has been computer vision for robotic systems, and this has allowed a degree of focus on some selected issues (albeit at the expense of some others) without significantly restricting the discussion. This approach has also, importantly, imparted a sense of real-world application to the discussion, even though the emphasis has been on basic principles rather than on the specific detail of any one particular problem that might have been selected.

Specifically, three identifiable perspectives on computer vision have been developed in the text, corresponding to the following:

1. the *transformation* of images,
2. the extraction of information to *describe* the contents of images,
3. the automatic *interpretation* of images.

Area 1 is equally important whether an image of interest is to be processed by a human being or by an artificially constructed system such as a computer; in either case it is useful to emphasize features of importance and, as far as possible, to maximize the quality of the image in question.

Area 2 can be of real significance in trying to quantify the features and characteristics of interest within an image and is, of course, of particular importance in many industrial applications of computer vision. Furthermore, the techniques developed within this area may generally be regarded as prerequisites for dealing in any meaningful way with questions arising in the interpretation of visual data.

Area 3 is, in many ways, the most demanding and challenging area of the three, and is perhaps the area where most rapid developments have taken place in recent years. It is clearly an area of great importance for the development of sensor-based robotic systems and, although this book has concentrated principally on relatively simple image structures, the foundations laid in these discussions are easily extended to more complex cases, particularly in situations where individual objects/items or sub-areas of interest can be located.

The introductory case study outlined in Chapter 1 in order to orient the subsequent discussions identified how these three areas of concern generate domain-specific problems to be tackled for a given application. Indeed, it should now be possible for the reader to return to that introductory example and specify in considerably more detail how the various problems raised there might be approached. To take just one example in each of the areas defined above:

Area 1 The extracted images representing alphanumeric characters may need to be processed using some form of smoothing algorithm in order to improve the quality of individual pattern samples for subsequent identification. (See, for example, the material of relevance in Chapters 2 and 3.)

Area 2 Character features, such as information about straight lines, closed boundaries, etc., may be used to produce a representation of the character classes to be identified. (See, for example, the material of relevance in Chapters 2, 3 and 4.)

Area 3 Features may be used to define a group of archetypal representations for each possible class, these being used as the basis for the implementation of some form of clustering procedure to identify individual samples subsequently. (See, for example, the material of relevance in Chapters 6, 7 and 8. Note also that in this case the particular approach described in Chapter 9 might be very applicable.)

Of course, other techniques might have been chosen by way of illustration and, indeed, many other factors would need to be taken into consideration (for example, the processing time required to execute an appropriate clustering algorithm might be significant here, and the variability in class-defining samples would need to be investigated, and so on), but a basic set of tools with which to tackle such practical problems has now been provided.

Interestingly, one area of study which some consider to be of primary importance to the continuing development of computer-based vision systems has hardly been mentioned here. This is the area of study concerned with vision in biological systems – arguably the most powerful and flexible visual processors known. A great deal can be learned about such areas as computational architectures, appropriate preprocessing, allocation of processing effort, and so on, by making a study of the mechanisms that determine the visual processing capabilities of biological organisms. Although this aspect of computer vision is largely neglected in the main body of the present text, its importance should not be overlooked and, for this reason, the Bibliography encompasses a modest number of references to the literature in this area.

In conclusion, then, it can be reiterated that computer vision, particularly in the context of sensor-driven robotic systems, is an extensive, expanding and hugely exciting field to study and, not surprisingly, one with a great deal of future potential. Current worldwide research programs, most notably those in the UK, the European Community, the USA and Japan, are giving increasing impetus to research in computer vision and, particularly, to the transfer of this technology from the research laboratories to its practical embodiment in systems of direct practical value in the industrial and commercial worlds.

This text has aimed to provide a base and a springboard from which the reader can begin to explore this fascinating subject, and the Bibliography provides pointers to more advanced and/or specialized areas for further investigation. This book will have achieved its aims if it has provided an accessible and coherent introduction to the various relevant sub-disciplines that need to be integrated for the specification, design and implementation of practical vision-based systems. Increasingly, such techniques are likely to play a part in encouraging more widespread utilization of robotic systems, and it is clear that, with increasing susceptibility to commercial pressures, vision systems are almost certain to have an impact in more and more areas. Coupling this observation with a recognition of the natural inquisitiveness of the human mind, it is clear that there may be more than a grain of truth in the suggestion that the current limitations of practical systems for computer vision will prove to be a challenge that future generations of scientists and engineers are unlikely to be able to resist.

Bibliography

ABDOU, I.E. & PRATT, W.K. 'Quantitative design and evaluation of enhancement/thresholding edge detection', *Proc. IEEE*, **67**, 1979, pp. 753–63.

AGIN, G.J. 'Computer vision systems for industrial inspection and assembly', *IEEE Comp. Mag.*, **13**, 1980, pp. 11–20.

AGRAWALA, A.K. (Ed.) *Machine recognition of patterns*, IEEE Press, 1977.

ALEKSANDER, I. 'Emergent intelligent properties of progressively structured pattern recognition nets', *Patt. Rec. Letts.*, **1**, 1983, pp. 375–84.

ALEKSANDER, I. *Designing intelligent systems: an introduction*, Kogan Page, 1984.

ANDREWS, H.C. *Introduction to mathematical techniques in pattern recognition*, Wiley, 1972.

ARBIB, M.A. *The metaphorical brain: an introduction to cybernetics as artificial intelligence and brain theory*, Wiley, 1972.

ARCELLI, C., CORDELLI, L. & LEVIALDI, S. 'Parallel thinning of binary pictures', *Electron. Letts.*, **11**, 1975, pp. 148–49.

ARGYLE, E. 'Techniques for edge detection', *Proc. IEEE*, **59**, 1971, pp. 285–87.

BALLARD, D.H. 'Generalising the Hough transform to detect arbitrary shapes', *Patt. Rec.*, **13**, 1981, pp. 111–22.

BALLARD, D.H. & BROWN, C.M. *Computer vision*, Prentice Hall, 1982.

BALLARD, D.H., HINTON, G.E. & SEJNOWSKI, T.J. 'Parallel visual computation', *Nature*, **306**, 1983, pp. 21–26.

BARROW, H. G. & TENENBAUM, J.M. 'Computational vision', *Proc. IEEE*, **69**, 1981, pp. 572–95.

BATCHELOR, B.G. *Practical approach to pattern classification*, Plenum Press, 1974.

BATCHELOR, B.G. (Ed.) *Pattern recognition: ideas in practice*, Plenum Press, 1978.

BATCHELOR, B.G. & MARLOW, B.K. 'Fast generation of chain code', *IEE Proc.*, **127** Part E, 1980, pp. 143–47.

BOW, S.-T. *Pattern recognition: applications to large data-set problems*, Marcel Dekker, 1984.

BRICE, C. & FENNEMA, C. 'Scene analysis using regions', *Artificial Intelligence*, **1**, 1970, pp. 205–26.

BURNS, J.B., HANSON, A.R. & RISEMAN, E.M. 'Extracting straight lines', *IEEE Trans. Patt. Anal. Machine Int.*, **8**, 1986, pp. 425–55.

CASTLEMAN, K.R. *Digital image processing*, Prentice Hall, 1979.

CHEN, C.H. *Statistical pattern recognition*, Spartan Books, 1973.

CHEN, C.H. (Ed.) *Pattern recognition and signal processing* (NATO Advanced Study Institute), Sijthoff & Noordhoff, 1978.

CHIN, R.T. & HARLOW, C.A. 'Automated visual inspection: a survey', *IEEE Trans. Patt. Anal. Machine Int.*, **4**, 1982, pp. 557–73.

COLEMAN, G.B. & ANDREWS, H.C. 'Image segmentation by clustering', *Proc. IEEE*, **67**, 1979, pp. 773–85.

DANIELSSON, P.-E. & LEVIALDI, S. 'Computer architectures for pictorial information systems', *IEEE Comp. Mag.*, **14**, 1981, pp. 53–67.

DASARATHY, B.V. & SHEELA, B.V. 'A composite classifier system design: concepts and methodology', *Proc. IEEE*, **67**, 1979, pp. 708–13.

DAVIES, E.R. & PLUMMER, A.P.N. 'Thinning algorithms: a critique and a new methodology', *Patt. Rec.*, **14**, 1981, pp. 53–63.

DAVIS, L.S. 'A survey of edge detection techniques', *Comp. Graphics Image Proc.*, **4**, 1975, pp. 248–70.

DEVIJVER, P.A. & KITTLER, J. *Pattern recognition: a statistical approach*, Prentice Hall, 1982.

DESSIMOZ, J.D. 'Specialised edge-trackers for contour extraction and line thinning', *Signal Processing*, **2**, 1980, pp. 71–73.

DESSIMOZ, J.D. & KAMMENOS, P. 'Software or hardware for robot vision', *Dig. Systems Indust. Automation*, **1**, 1982, pp. 143–60.

DODD, G.G. & ROSSOL, L. (Eds.) *Computer vision and sensor-based robots*, Plenum Press, 1979.

DOUGHERTY, E.R. & GIARDINA, C.R. *Matrix structured image processing*, Prentice-Hall, 1987.

DUDA, R.O. & HART, P.E. 'Use of the Hough transformation to detect lines and curves in pictures', *Commun. Ass. Comp. Mach.*, **15**, 1972, pp. 11–15.

DUDA, R.O. & HART, P.E. *Pattern classification and scene analysis*, Wiley, 1973.

DUDANI, S.A. & LUK, A.L. 'Locating straight line edge segments on outdoor scenes', *Patt. Rec.*, **10**, 1978, pp. 145–57.

DUFF, M.J.B. & LEVIALDI, S. (Eds) *Languages and architectures for image processing*, Academic Press, 1981.

DUFF, M.J.B. (Ed.) *Computing structures for image processing*, Academic Press, 1983.

DUFF, M.J.B. & FOUNTAIN, T.J. *Cellular logic image processing*, Academic Press, 1986.

FAIRHURST, M.C. & STONHAM, T.J. 'A classification system for alphanumeric characters based on learning network techniques', *Digital Processes*, **2**, 1976, pp. 321–40.

FAIRHURST, M.C. & KORMILO, S.R. 'Some economic considerations in the design of an optimal n-tuple pattern classifier', *Digital Processes*, **3**, 1977, pp. 321–30.

FAIRHURST, M.C. & MATTOSO MAIA, M. 'A two-layer memory network architecture for a pattern classifier', *Patt. Rec. Letts.*, **1**, 1983, pp. 267–71.

FAIRHURST, M.C. & MATTOSO MAIA, M. 'Performance comparisons in hierarchical architectures for memory-network pattern classifiers', *Patt. Rec. Letts.*, **4**, 1986, pp. 121–24.

FALLSIDE, F. & WOODS, W.A. (Eds) *Computer speech processing*, Prentice-Hall, 1985.

FOUNTAIN, T.J. & GOETCHERIAN, V. 'CLIP4 parallel processing system', *IEE Proc.*, **127** Part E, 1980, pp. 219–24.

FREEMAN, H. 'Computer processing of line drawing images', *Computer Surveys*, **6**, 1974, pp. 57–98.

FREI, W. & CHEN, C.C. 'Fast boundary detection: a generalisation and a new algorithm', *IEEE Trans. Comp.*, **26**, 1977, pp. 988–98.

FU, K.S. *Syntactic methods in pattern recognition*, Academic Press, 1974.

FU, K.S. & ROSENFELD, A. 'Pattern recognition and image processing', *IEEE Trans. Comp.*, **25**, 1976, pp. 1336–46.

FU, K.S. 'Recent developments in pattern recognition', *IEEE Trans. Comp.*, **29**, 1980, pp. 845–56.

FU, K.S. & YU, T.S. *Statistical pattern classification using contextual information*, Research Studies Press/Wiley, 1980.

FU, K.S (Ed.) *VLSI for pattern recognition and image processing*, Springer-Verlag, 1984.

FUKUNAGA, K. *Introduction to statistical pattern recognition*, Academic Press, 1972.

GLORIOSO, R.M. & OSORIO, F.C. *Engineering intelligent systems*, Digital Press, 1980.

GONZALEZ, R.C. & WINTZ, P. *Digital image processing*, Addison-Wesley, 1971.

GRANRATH, D.J. 'The role of human visual models in image processing', *Proc. IEEE*, **69**, 1981, pp. 552–61.

GROSKY, W.I. & JAIN, R. 'A pyramid-based approach to segmentation applied to region matching', *IEEE Trans. Patt. Anal. Machine Int.*, **8**, 1986, pp. 639–50.

GU, W.K. & HUANG, T.S. 'Connected line drawing extraction from a perspective view of a polyhedron', *IEEE Trans. Patt. Anal. Machine Int.*, **7**, 1985, pp. 422–31.

HAND, D.J. *Discrimination and classification*, Wiley, 1981.

HANSON, A.R. & RISEMAN, E.M. (Eds) *Computer vision systems*, Academic Press, 1978.

HILDITCH, C.J. 'Comparisons of thinning algorithms on a parallel processor', *Image Vision Comp.*, **1**, 1983, pp. 115–32.

HONG, T.H. & ROSENFELD, A. 'Compact region extraction using weighted pixel linking in a pyramid', *IEEE Trans. Patt. Anal. Machine Int.*, **6**, 1984, pp. 222–29.

HUBEL, D.H. & WIESEL, T.N. 'Brain mechanisms of vision', *Scientific American*, Sept. 1979, pp. 150–62.

HUECKEL, M.H. 'An operator which locates edges in digitized pictures', *Journal Ass. Comp. Mach.*, **18**, 1971, pp. 113–25.

IKEDA, K., HAMAMURA, T. *et al.* 'On-line recognition of hand-written characters utilising positional and stroke vector sequences', *Patt. Rec.*, **13**, 1981, pp. 191–206.

INIGO, R.M., McVEY, E.S. *et al.* 'Machine vision applied to vehicle guidance', *IEEE Trans. Patt. Anal. Machine Int.*, **6**, 1984, pp. 820–25.

INSTITUTION OF ELECTRICAL ENGINEERS, *Proc. 2nd. Int. Conf. on Image Processing and its Applications*, IEE Conference Publication No. 265, September 1986.

JAMES, M. *Classification algorithms*, Collins, 1985.

JARVIS, J.F. 'Visual inspection automation', *IEEE Comp. Mag.*, **13**, 1980, pp. 32–39.

KAHN, P. 'Local determination of a moving contrast edge', *IEEE Trans. Patt. Anal. Machine Int.*, **7**, 1985, pp. 402–10.

KANAL, L. 'Patterns in pattern recognition', *IEEE Trans. Info. Theory*, **20**, 1974, pp. 697–722.

KELLEY, R.B., MARTINS, H.A.S., *et al.* 'Three vision algorithms for acquiring workpieces from bins', *Proc. IEEE*, **71**, 1983, pp. 803–20.

KIMURA, E., TAKASHINA, K. *et al.* 'Modified quadratic discriminant functions and the application to Chinese character recognition', *IEEE Trans. Patt. Anal. Machine Int.*, **9**, 1987, pp. 149–53.

KITTLER, J. 'A method for determining k-nearest neighbours', *Kybernetes*, **7**, 1978, pp. 313–15.

KITTLER, J. 'Measurement errors in pattern recognition', *IEE Proc.*, **127** Part E, 1980, pp. 81–84.

KITTLER, J. & DUFF, M.J.B. (Eds) *Image processing system architectures*, Research Studies Press/Wiley, 1985.

KOHLER, R. 'A segmentation system based on thresholding', *Comp. Graphics Image Proc.*, **15**, 1981, pp. 319–38.

KOVALEVSKY, V.A. *Image pattern recognition*, Springer-Verlag, 1980.

KRUGER, R.P. & THOMPSON, W.B. 'A technical and economic assessment of computer vision for industrial inspection and robotic assembly', *Proc. IEEE*, **69**, 1981, pp. 1524–38.

KUSHNIR, M., ABE, K. & MATSUMOTO, K. 'An application of the Hough transform to the recognition of printed Hebrew characters', *Patt. Rec.*, **16**, 1983, pp. 183–92.

LEVINE, M.D. *Vision in man and machine*, McGraw-Hill, 1985.

LIPKIN, B.S. & ROSENFELD, A. (Eds) *Picture processing and psychopictorics*, Academic Press, 1970.

MANTAS, J. 'An overview of character recognition methodologies', *Patt. Rec.*, **19**, 1986, pp. 425–30.

MANTAS, J. 'Methodologies in pattern recognition and image analysis – a brief survey', *Patt. Rec.*, **20**, 1987, pp. 1–6.

MARR, D. *Vision*, Freeman, 1982.

McDONALD, A.C. *Robot technology: theory, design and applications*, Prentice Hall, 1986.

MENDEL, J.M. & FU, K.S. (Eds) *Adaptive, learning and pattern recognition systems: theory and applications*, Academic Press, 1970.

MORI, S., YAMAMOTO, K. & YASUDA, M. 'Research on machine recognition of hand-printed characters', *IEEE Trans. Patt. Anal. Machine Int.*, **6**, 1984, pp. 386–405.

MYERS, W. 'Industry begins to use visual pattern recognition', *IEEE Comp. Mag.*, **13**, 1980, pp. 21–31.

NAGY, G. 'State of the art in pattern recognition', *Proc. IEEE*, **56**, 1968, pp. 836–62.

NALWA, V.S. & BINFORD, T.O. 'On detecting edges', *IEEE Trans. Patt. Anal. Machine Int.*, **6**, 1986, pp. 699–714.

NARAYANAN, K.A. & ROSENFELD, A. 'Image smoothing by local use of global information', *IEEE Trans. Systems, Man Cybernetics*, **11**, 1981, pp. 826–31.

NEVATIA, R. *Machine perception*, Prentice Hall, 1982.

NI, L.M. & JAIN, A.K. 'A VLSI systolic architecture for pattern clustering', *IEEE Trans. Patt. Anal. Machine Int.*, **7**, 1985, pp. 80–90.

NIBLACK, W. & MATSUTAMA, T. 'Edge preserving smoothing', *Comp. Graphics Image Proc.*, **9**, 1979, pp. 394–407.

NIBLACK, W. *An introduction to digital image processing*, Prentice Hall, 1986.

NITZAN, D. & ROSEN, C.A. 'Programmable industrial automation', *IEEE Trans. Comp.*, **25**, 1976, pp. 1259–70.

N-NAGY, F. *Engineering foundations of robotics*, Prentice Hall, 1987.

OFFEN, R.J. (Ed.) *VLSI image processing*, Collins, 1985.

PAL, S.K. & PAL, N.R. 'Segmentation using contrast and homogeneity measures', *Patt. Rec. Letts.*, **5**, 1987, pp. 293–304.

PALIWAL, K.K. & RAO, P.V.S. 'Application of *k*-nearest neighbour decision rule in vowel recognition', *IEEE Trans. Patt. Anal. Machine Int.*, **5**, 1983, pp. 229–31.

PAVLIDIS, T. 'A review of algorithms for shape analysis', *Comp. Graphics Image Proc.*, **7**, 1978, pp. 243–58.

PAVLIDIS, T. *Algorithms for graphics and image processing*, Computer Science Press, 1982.

PELI, T. & MALAH, D. 'A study of edge detection algorithms', *Comp. Graphics Image Proc.*, **20**, 1982, pp. 1–21.

PERKINS, W. A. 'INSPECTOR: a computer vision system that learns to inspect parts', *IEEE Trans. Patt. Anal. Machine Int.*, **5**, 1983, pp. 584–92.

PORTER, G.B. & MUNDY, J.L. 'Visual inspection system design', *IEEE Comp. Mag.*, **13**, 1980, pp. 40–49.

POT, J., COIFFET, P. & RIVES, P. 'Comparison of five methods for the recognition of industrial parts'. *Dig. Systems Indust. Automation*, **1**, 1982, pp. 289–303.

PRATT, W.K. *Digital image processing*, Wiley, 1978.

PRESTON, K. 'A comparison of analog and digital techniques for pattern recognition', *Proc. IEEE*, **60**, 1972, pp. 1216–31.

PRESTON, K. & ONOE, M. (Eds) *Digital processing of biomedical images*, Plenum Press, 1976.

PRESTON, K. 'Gray level image processing by cellular logic transforms', *IEEE Trans. Patt. Anal. Machine Int.*, **5**, 1983, pp. 55–58.

PUGH, A. (Ed) *Robot vision*, Springer-Verlag, 1983.

RANKY, P. *Computer integrated manufacture: an introduction with case studies*, Prentice-Hall, 1986.

REDDY, R. 'Robotics and the factory of the future', Editorial in *Proc. IEEE*, **71**, 1983, pp. 787–88.

ROSENFELD, A. & THURSTON, M. 'Edge and curve detection for visual scene analysis', *IEEE Trans. Comp.*, **20**, 1971, pp. 562–69.

ROSENFELD, A. & KAK, A.C. *Digital picture processing*, Academic Press, 1976.

ROSENFELD, A. 'Image pattern recognition', *Proc. IEEE*, **69**, 1981, pp. 596–605.

SANDERSON, A.C. & PERRY, G. 'Sensor-based robotic assembly systems: research and applications in electronic manufacturing', *Proc. IEEE*, **71**, 1983, pp. 856–71.

SCALTOCK, J. 'A survey of the literature of cluster analysis', *Comp. Journal*, **25**, 1982, pp. 130–33.

SCHOLTEN, D.K. & WILSON, S.G. 'Chain coding with a hexagonal lattice', *IEEE Trans. Patt. Anal. Machine Int.*, **5**, 1983, pp. 526–33.

SETHI, I.K. 'A fast algorithm for recognising nearest neighbours', *IEEE Trans. Systems, Man Cybernetics*, **11**, 1981, pp. 245–48.

SHNEIER, M. 'Using pyramids to define local thresholds for blob detection', *IEEE Trans. Patt. Anal. Machine Int.*, **5**, 1983, pp. 345–49.

SKLANSKY, J. 'Image segmentation and feature extraction', *IEEE Trans. Syst. Man Cybernetics*, **8**, 1978, pp. 237–47.

SMITH, R.W. 'Computer processing of line images: a survey', *Patt. Rec.*, **20**, 1987, pp. 7–15.

SNYDER, W.E. & COWART, A. 'An iterative approach to region growing using associative memories', *IEEE Trans. Patt. Anal. Machine Int.*, **5**, 1983, pp. 349–52.

SNYDER, W.E. *Industrial robots: computer interfacing and control*, Prentice Hall, 1985.

SOMMERHOFF, G. *Logic of the living brain*, Wiley, 1974.

SPOEHR, K.T. & LEHMKUHLE, S.W. *Visual information processing*, Freeman, 1982.

SURESH, B.R., FUNDAKOWSKI, R.A. *et al.* 'A real-time automated visual inspection system for hot steel slabs', *IEEE Trans. Patt. Anal. Machine Int.*, **5**, 1983, pp. 563–72.

TANIMOTO, S. & PAVLIDIS, T. 'A hierarchical data structure for picture processing', *Comp. Graphics Image Proc.*, **4**, 1975, pp. 104–19.

TANIMOTO, S. & KLINGER, A. *Structured computer vision*, Academic Press, 1980.

TOU, J.T. & GONZALEZ, R.S. *Pattern recognition principles*, Addison-Wesley, 1974.

TURNEY, J.L., MUDGE, T.N. & VOZZ, R.A. 'Recognising partially occluded parts', *IEEE Trans. Patt. Anal. Machine Int.*, **7**, 1985, pp. 410–22.

ULLMANN, J.R. *Pattern recognition techniques*, Butterworths, 1973.

ULLMANN, J.R. & LAFFERTY, H.H. 'Picture-processing techniques for geometric tolerance checking of integrated-circuit layouts', *IEE Proc.*, **127** Part E, 1980, pp. 8–17.

WECHSLER, H. 'A new and fast algorithm for estimating the perimeter of objects for industrial vision tasks', *Comp. Graphics Image Proc.*, **17**, 1981, pp. 375–81.

WINSTON, P.H. (Ed.) *The psychology of computer vision*, McGraw-Hill, 1975.

YEH, P.-S., ANTOY, S. *et al.* 'Address location on envelopes', *Patt. Rec.*, **20**, 1987, pp. 213–28.

YOUNG, T.Y. & CALVERT, T.W. *Classification, estimation and pattern recognition*, Elsevier, 1974.

Index